SPRINGER PROTOCOLS HANDBOOKS

For further volumes:
http://www.springer.com/series/8623

Springer Protocols Handbooks collects a diverse range of step-by-step laboratory methods and protocols from across the life and biomedical sciences. Each protocol is provided in the Springer Protocol format: readily-reproducible in a step-by-step fashion. Each protocol opens with an introductory overview, a list of the materials and reagents needed to complete the experiment, and is followed by a detailed procedure supported by a helpful notes section offering tips and tricks of the trade as well as troubleshooting advice. With a focus on large comprehensive protocol collections and an international authorship, Springer Protocols Handbooks are a valuable addition to the laboratory.

Plant-Microbe Interactions

Laboratory Techniques

M. Senthilkumar

Division of Basic Sciences, Indian Institute of Pulses Research, Kanpur, Uttar Pradesh, India

N. Amaresan

C.G. Bhakta Institute of Biotechnology, Uka Tarsadia University, Surat, Gujarat, India

A. Sankaranarayanan

C.G. Bhakta Institute of Biotechnology, Uka Tarsadia University, Surat, Gujarat, India

Humana Press

M. Senthilkumar
Division of Basic Sciences
Indian Institute of Pulses Research
Kanpur, Uttar Pradesh, India

N. Amaresan
C.G. Bhakta Institute of Biotechnology
Uka Tarsadia University
Surat, Gujarat, India

A. Sankaranarayanan
C.G. Bhakta Institute of Biotechnology
Uka Tarsadia University
Surat, Gujarat, India

ISSN 1949-2448 ISSN 1949-2456 (electronic)
Springer Protocols Handbooks
ISBN 978-1-0716-1079-4 ISBN 978-1-0716-1080-0 (eBook)
https://doi.org/10.1007/978-1-0716-1080-0

This Humana imprint is published by the registered company Springer Science+Business Media, LLC, part of Springer Nature.
The registered company address is: 1 New York Plaza, New York, NY 10004, U.S.A.

Preface

The world is much concerned about meeting the food demand of emerging population in the context of various biotic and abiotic stresses affecting the growth of the plants, yield, and sustainable agricultural development. Agricultural research to improve and sustain food production especially using beneficial plant–microbe interactions is gaining importance. Important research protocols have been presented in *Plant-Microbe Interactions: Laboratory Techniques* for the benefit of global researchers and student community.

This manual comprises of 83 protocols given under five major sections: biological nitrogen fixation, mineral solubilization and mobilization, abiotic stress management, biocontrol of plant pathogens, and biological control of insects and nematodes. In general, this manual covers various protocols relating to plant–microbe interactions with the preparation of reagents and important instructions/special note relating to the practicals. We devoted much care to the preparation and compiling of each and every protocol provided in the manual.

We strongly believe that the lab manual will meet the practical and research needs of postgraduate students, research scholars, postdoctoral fellows, and scientists as well as teachers from various fields, including pathology, microbiology, entomology, and agronomy. We thank the **Springer Publications** for their support and cooperation for bringing out this lab manual in a nice manner and on time. We highly appreciate your valuable inputs and concrete suggestions for the improvement of the manual.

Kanpur, Uttar Pradesh, India — *M. Senthilkumar*
Surat, Gujarat, India — *N. Amaresan*
Surat, Gujarat, India — *A. Sankaranarayanan*

About the Authors

M. Senthilkumar is currently a principal scientist (microbiology) at ICAR-Indian Institute of Pulses Research, Kanpur. He is a postgraduate of agricultural microbiology and holds a PhD (Microbiology) from the Indian Agricultural Research Institute, New Delhi. He received postdoctoral fellowship from the Korean Atomic Energy Research Institute, South Korea. He has 15 years of research experience in various areas of plant–microbe interactions, including biological control of soybean diseases and microbial-mediated abiotic stress management in pulses. He also explored the diversity of endophytic bacteria from wild and cultivated soybean varieties for biological control of soybean charcoal rot disease. He has handled numerous research projects funded, e.g., by the Department of Biotechnology (DBT), Indian Council of Agricultural Research (ICAR). He has developed bio-inoculants and microbial consortia for improving moisture-deficit stress tolerance of rain-fed chickpea. He has published more than 20 research papers in peer-reviewed journals. He is a fellow of the Indian Society for Pulses Research and Development (ISPRD), Kanpur, India.

N. Amaresan is an assistant professor at C.G. Bhakta Institute of Biotechnology, Uka Tarsadia University, Gujarat. He holds undergraduate, postgraduate, and doctorate degrees in microbiology. Dr. N. Amaresan has over 14 years of experience in teaching and research in various allied fields of microbiology, including microbial ecology and plant–microbe interactions. He has been awarded Young Scientist Awards by the Association of Microbiologists of India, National Academy of Biological Sciences, and is the recipient of a visiting scientist fellowship from the National Academy of India. He has published more than 60 research articles, books, and book chapters. He has handled various projects sponsored by institutions such as DBT, DST, and GEMI.

A. Sankaranarayanan is associated with C.G. Bhakta Institute of Biotechnology, Uka Tarsadia University, Gujarat from 2015 onwards. His current research focus is on antibacterial potential of herbal products against multidrug-resistant microbial pathogens, bacterial diversity in fermented batter, microbial spoilage of fermented food, and impact of irradiation on fermented food. He has received the Indian Academy of Sciences–sponsored summer research fellowship for young teachers category for 3 years consecutively. He has published books, book chapters, research articles in international and national journals, guided PhD and MPhil scholars, and operated funded projects. His name is included as a Mentor in DST-Mentors/Resource Persons for summer/winter camps and other INSPIRE initiatives, Department of Science & Technology, Government of India, New Delhi. He is a Grant reviewer in British Society of Antimicrobial Chemotherapy (BSAC), UK.

Part I

Techniques for Studying Biological Nitrogen Fixation

Check for updates

Chapter 1

Isolation, Characterization, and Preservation of Rhizobia from Root Nodules of Legumes

Abstract

Rhizobium initiates root nodules in legume plants and fix atmospheric nitrogen symbiotically. Rhizobia is the collective term used to indicate a bacterium that has the ability to initiate root nodulation in legumes and fix nitrogen symbiotically. They are heterogeneous assemblage of gram-negative, aerobic, non-sporulating rod shaped, symbiont of legumes, and soil bacteria. Only about 20% of the total of about 20,000 species and 57% of about 750 genera of legume plants has been studied for nodulation. The taxonomy and diversity of rhizobia were changed enormously in last two decades. Currently rhizobia consist of 176 species spread over 15 genera and have been validated using molecular markers. There are many legumes and ecosystems that have not been analyzed yet and then the number of rhizobial species and symbiovars are expected to increase in further. This chapter deals with collection and preservation of root nodules, isolation, characterization of rhizobia and their preservation methods.

Keywords Root nodules, Rhizobium, Diversity, YEMA, Pure culture, Characterization, Preservation

1 Isolation of Rhizobia from Root Nodules of Legumes

1.1 Materials

1. Healthy legume plants
2. Polythene bags
3. Sterile microcentrifuge tubes or vials
4. Mercuric chloride (0.01%)
5. Sodium hypochlorite (3%)
6. Ethyl alcohol (70%)
7. Glycerol (15%)
8. Nutrient Agar—petri plates
9. Yeast Extract Mannitol Agar (YEMA) medium
10. Sterile pestle and mortar
11. Sterile water

M. Senthilkumar, N. Amaresan and A. Sankaranarayanan, *Plant-Microbe Interactions: Laboratory Techniques*, Springer Protocols Handbooks, https://doi.org/10.1007/978-1-0716-1080-0_1, © Springer Science+Business Media, LLC, part of Springer Nature 2021

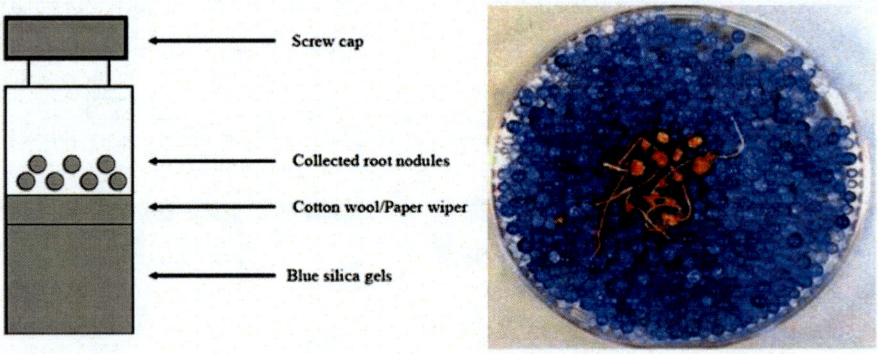

Fig. 1 Preservation of root nodules in silica gel

1.2 Methods

1.2.1 Isolation of Rhizobia from Preserved Root Nodules

1. Healthy legume plants should be uprooted with the help of shovels and carefully collect soils around root nodules in small plastic Ziplocs bags for characterization.
2. Exposed root nodules can be collected with forceps and place them in a vial with desiccated silica gels shown in Fig. 1. If the gel color turns pink, the desiccant must be changed for the blue one as soon as possible to preserve the nodules. For field trips of more than 2 days, the root nodules collected must be prevented from decomposing and invasion by other soil microorganisms which interfere with subsequent isolation procedures.
3. All nodules from a single host plant represent one unit of collected material and store them in the same vial while root nodules from different plants of the same species should not be combined. Keep the vials with dried root nodules in a refrigerator at 4 °C until the isolation of rhizobia.
4. Wash dried root nodules under tap water to remove soil contamination and transfer to sterile water. Keep the root nodules in sterile water in a refrigerator at 4 °C overnight to absorb water.
5. Sterilize the surfaces of root nodules by placing them in 70% ethanol for 30 s, and in a 3% sodium hypochlorite solution for 3 min. Subsequently wash the root nodules at least five times with sterile water. Crush the root nodule in a microfuge tube with 100 μL of sterile glycerol (15%) solution.
6. Take a loopful of crushed cell suspension and streak-inoculate onto the surface of a yeast extract mannitol agar (YMA) medium [1]. The turbid suspensions with 15% glycerol solution in a microfuge tube should be kept in freezer at −30 °C until isolations of rhizobia are complete. If first isolations do not succeed, the glycerol solutions are re-streaked onto a fresh YEMA plate to obtain bacterial growth.

7. Incubate the plates at 28 °C for 2 weeks in an inverted position. Re-streak the well-separated single colonies onto fresh plate to obtain pure cultures.

1.2.2 Isolation of Rhizobia from Fresh Root Nodules

1. Uproot the healthy legume plants. Transfer the whole plant with root system with soil to the laboratory in a polythene bag to avoid moisture loss.
2. Wash their root system with tap water to remove soil particles. Detach the root nodules which are healthy, pink and preferable on the tap root with the help of forceps, wash it with sterile water and transfer into sterile microcentrifuge tubes/vials with appropriate size for surface sterilization.
3. Add enough volume (10 volumes of nodule sample) of mercuric chloride 0.01%, mix thoroughly for 3 min and decant the used mercuric chloride.
4. Add 70% ethyl alcohol for 30 s. Decant the used alcohol and wash the surface-sterilized nodules with sterile water for at least seven times in order to eliminate traces of alcohol and mercuric chloride. Spread-inoculate the water (100 μL) from the final washes on Nutrient Agar—petri plates for checking surface sterilization.
5. Transfer the surface-sterilized root nodules to a sterile pestle and mortar and then macerates with known quantity of sterile water.
6. Serially dilute the nodule cell macerate upto 10^{-4} with sterile water and streak-/spread-inoculate onto petri plates containing yeast extract mannitol agar (YEMA) medium [2] and then incubate at 28 °C for 7 days.

1.3 Observations

1. Monitor the appearance of bacterial colonies along with their morphological traits on YEMA at every 24 h. Purify the translucent rhizobia like colonies through frequent streaking on fresh YEMA plates and then transfer every single pure colonies into 25% glycerol stock for storage under −20 °C deep freezer.

1.4 Precautions

1. **Site identification:** Record the details on country, prefecture, nearest town, kilometers from the nearest town, latitude, and longitude by using GPS.
2. **Host plants identification:** Collect a sample of the legume plant for herbarium specimen to identify genus, species, and cultivar. Take pictures of the leguminous plants. Consult a specialist in legume plant taxonomy.
3. In case of leguminous crops such as soybean and mung bean grown in farmers' fields, ask the farmers the variety of the leguminous crop and confirm history of the fields concerning inoculation of rhizobia.

4. The needle method of isolation is especially useful with freshly harvested nodules 2 mm or larger in diameter. Wash the nodule initially with water, alcohol, and then held with forceps and briefly passing the nodule through a flame. Place the surface-sterilized nodule on a small piece of sterile filter paper (2 × 2 cm) in a sterile petri dish. A new piece of filter paper should be used for each nodule. The same petri dish can be used for several nodules. Dip the blunt-tipped forceps into 95% alcohol and flamed momentarily. While holding the nodule with the forceps and resting the nodule on sterile filter paper, slice a small section with a flamed, hot scalpel. While still holding the nodule with the forceps on the filter paper, insert the tip of a sterile inoculation needle (with a 1-mm loop) into the cut surface. Streak the loaded inoculum in the needle directly onto a YMA plate containing Congo red (CR) and a YMA plate containing BTB.
5. Sterility check: Bacterial colonies isolated from nodules whose washouts after sterilization has no bacterial contamination should be considered for further characterization.
6. Slow/fast-growing nature of rhizobia: As the root nodulating bacteria are diverse in nature for growth rate and colony morphology, it is essential to monitor the bacterial growth for longer incubation time.

2 Characterization of Rhizobia

2.1 Presumptive Tests

1. **Growth on the peptone–glucose agar medium:** Prepare the peptone–glucose agar medium by dissolving peptone 10.0 g, glucose 5.0 g, agar 15.0 g, and 10 mL of bromocresol purple in 1000 mL of distilled water and adjust the pH to 6.8. Streak-inoculate the *Rhizobium* isolates onto peptone–glucose agar medium. Incubate the petri plates at 28 ± 2 °C for 5 days to observe the growth and change in the color.
2. **Growth on YEMA with BTB:** Inoculate the YEM medium enriched with BTB @ (25 μg/mL) to selectively identify rhizobium. Freshly prepared YMA plates containing bromothymol blue have a pH of 6.8 and are green. Incubate and make daily observations for the appearance of colonies typical of rhizobia. Colonies should show little or no Congo red absorption when incubated in the dark. A blue color indicative of an alkaline reaction on BTB should be obtained with slow-growing *Bradyrhizobium* spp. A yellow color (acid) reaction is usually produced by the fast-growing *Rhizobium* spp. Plates should be read for reactions after 3–5 days (fast-growers, e.g., bean rhizobia) and 5–7 (slow-growers, e.g., soybean rhizobia). Rhizobia generally do not absorb Congo red when plates are

incubated in the dark. Colonies remain white, opaque, or occasionally pink. Contaminating organisms usually absorb the red dye. However, reactions depend on the concentration of Congo red and age of the culture. Rhizobia will absorb the red dye if plates are exposed to light during the incubation or exposed to light for an hour or more after growth has occurred.

2.2 Confirmative Plant Infection Test

For large seeded legumes like beans, soybean, groundnut, and cowpea, Leonard jars and growth-pouches are recommended as growth units for authentication. Alternatively, smaller seeded siratro (*Macroptilium atropurpureum*) are grown in nitrogen free plant nutrient-agar slants. Siratro is used in authenticating most bradyrhizobia from tropical legumes because it nodulates with more than 90% of all bradyrhizobia including those associated with cowpea, groundnut, and promiscuous soybean.

1. Set up 2–5 suitable growth units for each of the isolates plus at least two extra units that will serve as uninoculated controls.
2. Surface-sterilize and pre-germinate the seeds of host plants.
3. Inoculate 1 mL of broth culture for each isolate onto each of the pre-germinated seeds in two growth units.
4. Examine plants for differences in vigor and color between the inoculated and uninoculated at 15–30 days of growth.
5. Remove the plants from the rooting medium and note the presence or absence of nodules. The presence of nodules in the non-inoculated treatment invalidates the test. If the PI tests are satisfactory, the isolates are regarded as fully authenticated cultures. The cultures of presumptive isolates are now confirmed as rhizobia and may be given collection numbers.

3 Preservation of Rhizobia

There are a number of satisfactory methods for preserving rhizobial cultures including yeast mannitol agar (YMA) slants in screw-cap tubes, desiccation on porcelain beads, lyophilized (freeze-dried), and as frozen liquid suspension under liquid nitrogen. YMA slants and porcelain beads are recommended for laboratories with limited resources.

1. To prepare for storage on beads, inoculate a loopful of culture from YMA slant into 3 mL of sterile YM-broth and incubate to maximum turbidity on a rotary shaker. Place 20–30 ceramic beads (washed and oven dried) in a screw-cap test tube, cover the mouth of the tube with foil, and sterilize in the oven for 1–2 h at 160–170 °C.

2. Prepare storage tubes using 6–7 g silica gel and sufficient cotton or glass wool to keep the silica gel in place. The rubber lined caps for the tubes must be autoclaved separately and then dried in an oven at 80–90 °C.
3. The glass wool may be oven sterilized in the storage tube with the silica gel. When cotton is used, it should be autoclaved in small balls in a foil covered beaker. These cotton balls should be of a suitable size to facilitate easy aseptic transfer to the storage tube with forceps. Residual moisture is removed in the oven at 70–80 °C before transferring it aseptically to the sterile storage tubes. The autoclaved caps are then added to the tubes.
4. Transfer the sterilized beads aseptically to the broth culture in the tubes and replug. Soak the beads for 1–2 h, then invert the tube and allow the excess broth culture to soak into the cotton plug.
5. Transfer the beads impregnated with rhizobia into the storage tube aseptically, replace and tighten the screw caps securely. Examine the tubes after a day or so to ensure that the silica gel is still blue. If it turns pink or colorless, then too much moisture was transferred with the beads or an improper seal is permitting entry of moisture.
6. To regenerate a culture, inoculate YM-broth with one or two beads. These are easily speared from the storage tube using a sterile needle with a slight hook. Once the broth becomes turbid, loopfuls should be streaked on presumptive test media to check for purity. Subculture from the broth onto YMA slants as desired.

References

1. Somasegaran P, Hoben HJ (1994) Handbook for Rhizobia. Methods in legume-*Rhizobium* technology. Springer, New York, NY, pp 332–341
2. Vincent JM (1970) A manual for the practical study of root-nodule bacteria. In: International biological program handbook. Blackwell Scientific Publication, Oxford

Chapter 2

Enumeration of Rhizobia: Most-Probable-Number (MPN) Technique

Abstract

Rhizobial diversity and abundance varies with different agroecosystems depending on its physicochemical properties. Since oxygen is one of the key factors that regulate microbial activity, most of the rice fallow soils showed either lack of native rhizobia or their low abundance, at about 100 rhizobia/g of dry soil. It is essential to enumerate the rhizobial populations in soil to determine the need of artificial inoculation of rhizobia. Since the absence of significant molecular markers to differentiate rhizobia from rest of the soil bacteria, most-probable-number (MPN) techniques based on observation of presence or absence of root nodule is used to enumerate rhizobial population. Rhizobial MPN enumeration is based on the assumptions that a single, viable rhizobial cell in the root area of a young legume host of the appropriate type will cause nodule formation in nitrogen free media.

Keywords *Rhizobium*, *Enumeration*, Most-probable-number (MPN), Plant-infection assays, Range of transition (ROT)

Most-probable-number (MPN) technique is a means to estimate microbial population sizes. The technique is widely used to enumerate rhizobia based upon the ability of rhizobia to nodulate appropriate host legume plants. The method relies upon the pattern of positive and negative nodulation responses of host plants inoculated with a consecutive series of dilutions of rhizobia containing sample suspension. All rhizobial MPN plant-infection assays are similar in principle, but vary widely in design and implementation. Any MPN assay of a legume inoculant is based on an initial series of dilutions of the inoculant, which is then further serially diluted, and the second series of dilutions (dilution ratio) inoculated in successive fashion onto a number of replicate plant growth units. Generally, the higher the number of plant replicates used and the lower the base dilution ratio, the smaller will be the 95% confidence limits on either side of the actual population estimate obtained [1].

Rhizobial MPN enumeration is based on the following major assumptions:

M. Senthilkumar, N. Amaresan and A. Sankaranarayanan, *Plant-Microbe Interactions: Laboratory Techniques*, Springer Protocols Handbooks, https://doi.org/10.1007/978-1-0716-1080-0_2, © Springer Science+Business Media, LLC, part of Springer Nature 2021

- A single, viable rhizobial cell in the root area of a young legume host of the appropriate type will cause nodule formation in nitrogen free media.
- Nodule formation is evidence of the presence of viable infective rhizobia.
- The validity of the MPN assay is demonstrated by the absence of nodules on uninoculated plants that are otherwise treated identically to the inoculated plants.
- Absence of nodule formation is evidence of the absence of viable infective rhizobia.
- Rhizobia present in the initial and all subsequent dilutions are evenly distributed.

The process of designing an MPN assay for legume inoculants involves selecting:

- the degree of initial dilution of the sample
- the base dilution ratio of the serial dilutions applied to plants
- the number of serial dilution steps applied to plants
- the number of replicate plants to be inoculated at each dilution level
- the volume of inoculant applied to each plant

Standard MPN tables do not provide estimates for all dilution ratio—replicate combinations. The upper range of rhizobial populations which can be detected with an MPN are functions of (1) the amount of dilution provided to the sample prior to the serial dilutions applied to the plants, (2) the base dilution ratio of dilutions applied to the plants, and (3) the number of dilution steps. If the number of rhizobia in an inoculant cannot be anticipated, a very wide range can be accommodated by using a tenfold dilution series. However, fivefold or fourfold dilution ratios result in greater accuracy.

1 Method

1. Using growth pouches and a suitable supporting rack prepare 30 pouches (growth units) with a suitable number of healthy host plants per pouch and sterile N-free nutrient solution (30 mL for whole pouches, 15 mL per side for split pouches). This makes six sets of five growth units.
2. Taking a representative sample of the inoculant, dilute the sample by successive tenfold dilutions (in sterile diluent) to a suitable starting point based on the anticipated number of rhizobia in the product. Given an anticipated rhizobial cell

number of greater than 10^9 per g, a suitable starting point for further dilutions would be a dilution of 10^{-6}.

3. From the 10^{-6} dilution make an additional six serial fivefold dilutions in sterile diluent (the initial dilution is 10^{-6}; the base dilution ratio is 5).
4. Apply 1.0 mL to each of four replicate growth units from each of the six fivefold dilutions. Leave the fifth growth unit in each series uninoculated. Whether one works from the most dilute of the fivefold dilutions to the least dilute, or vice versa, depends upon the technique used to apply the inoculant suspension, but in any case work in order, and be consistent.
5. Place the growth-unit rack containing the MPN assay in an appropriate growth environment and maintain suitable plant condition by watering with sterile water.
6. After 3–4 weeks (depending upon the legume host used) examine the plants and record each growth pouch unit as "+" or "−" for nodulation. Even one nodule per growth unit means a "+" score for that unit. All six negative control units must be free of nodules (if not, discard the test, repeat, and work on technique).
7. For each of the six successive fivefold dilutions a number of pouches between 0 and 4 will have been scored as "+" for nodulation. The number of "+" scores taken in order of increasing dilution will yield a six digit number. A typical result could be: 4, 4, 4, 1, 1, 0.
8. Using an MPN probability table based on six fivefold dilutions, four replicates, and 1.0 mL volume applied per growth unit, determine the population estimate, and 95% confidence limits. This estimate refers to the number of rhizobia present in 1 mL of the suspension from which the first fivefold dilution applied to the plants was made. Multiply the MPN estimate by the reciprocal of the dilution level of the inoculant suspension prior to beginning the fivefold serial dilutions (in this case 10^6) to get an estimate for the number of viable rhizobia in the original inoculant. For the example given above the estimate is $379 \times 10^6 = 3.79 \times 10^8$ rhizobia per g.

2 Precautions

1. **Selection of a plant growth system:** The success of MPN plant-infection techniques is largely dependent upon the ability of the researcher to produce healthy, uniform host plants. These assays are best conducted in a growth chamber or growth room; however, some plant systems are suitable for the greenhouse. Attempts to grow the host legumes next to a

"sunny window" seldom provide the plants with sufficient light, and should not be attempted. Determine the type and number of growth containers needed. Prepare enough to discard the least uniform 25% of the plants. The units must be completely free of rhizobia or, preferably, sterile.

2. **Plant containers for host legumes:** Size of the plant growth container is determined by the size of the host legume seed, the growth rate of the plant, and the number of seeds planted in each container. A container of the proper size will allow the roots to completely explore the media and will enhance the probability that the roots will contact the rhizobia. The likelihood of outside contamination is also a consideration when selecting the container.

   Growth pouches—Growth pouches are appropriate for both small and large seeded legumes such as: *Medicago sativa*, *Trifolium pratense*, *Onobrychis viciaefoli*, *Glycine max*, *Arachis hypogaea*, *Phaseolus vulgaris*, *Phaseolus lunatus*, and *Vigna unguiculata*. The pouches consist of a semirigid plastic bag containing a flat paper wick. Thirty milliliter of N-free plant nutrient solution is placed inside the pouch. The seeds are planted in a trough at the top of the wick. The pouches are normally initially free of rhizobia, and can usually be used in plant-infection assays without prior sterilization. A flexible plastic straw (or disposable Pasteur pipette) can be inserted into the pouches to facilitate watering. This decreases the risk of cross-contamination that is associated with the watering process. Insects can also be a source of contamination when using pouches. Partially closing the pouches with tape, paper clips, or staples can help to reduce contamination. A simple wire rack can be constructed to hold the growth pouches.

   Agar slants—Small-seeded legumes can also be grown in foam (or cotton plug) stoppered glass tubes containing N-free plant nutrient agar slants. Nodulation of uninoculated plants is rare, due in large part to the protection afforded by the stopper at the mouth of the tube.

   Leonard jars—While Leonard jars are commonly used as aseptic growth containers in the study of the legume—*Rhizobium* symbiosis, the value of Leonard jars and similar systems using vermiculite is questionable for plant-infection MPN counts for several reasons. Limitations include the inability to apply the inoculant directly to root surfaces, the difficulties in observing and recovering nodules, and the possibility of the vermiculite preventing contact of the rhizobia with the roots. These disadvantages frequently result in statistically unacceptable MPN results and underestimates of the population size.

3. **Preparation of plant nutrient solution:** Two different plant nutrient solutions which have been used successfully in MPN

assays are prepared as shown in Table 1. It is convenient to prepare stock solutions for use in preparation of nutrient solutions. When preparing the nutrient solution, it is important to dissolve the two chemicals for stock number 6 separately before combining. If nutrient solution is to be autoclaved, the $CaSO_4$ should be sterilized separately and added last and after the autoclaving to prevent formation of an insoluble precipitate.

4. **Preparation and germination of seed and establishment of plants in growth units:** Determine the amount of seed needed based on germination percentage, surface-sterilize and pre-germinate the seeds. Many legumes have impermeable seed coats that require scarification for uniform germination. These seeds can be treated with concentrated sulfuric acid, which simultaneously surface sterilizes and scarifies the seed (Table 2). Wash the seeds at least five times in sterile water following surface sterilization or acid scarification. Small-seeded legumes can be sown unimbibed in prepared growth pouches and allowed to germinate in darkness at 20 °C. If the growth units chosen for small-seeded legumes are agar slants, the seed should be pre-germinated on water agar (1%) in petri plates (temperature 20–30 °C depending on the host legume; temperate, tropical, etc.). Incubate the plates inverted so that the radicles do not grow into the agar. When sowing small-seeded legumes in agar slant tubes, the germinated seeds are individually transferred to the slant when the radicles are 0.2–1.0 cm in length. The radicles should be placed on the surface of the slant and not embedded into the media. For large seeded legumes sown in growth pouches, it is necessary to refold the wick so that the paper trough is deeper to prevent seed dehydration and to poke holes in the trough of the pouch and insert the radicle of each seed into a hole. Grow the plants for 3–6 days before inoculation, and discard growth units showing significant lack of uniformity in the number of plants or in root or shoot development.

5. **Preparing dilutions and inoculating the host plants:** The dilution procedure must be a systematic and accurate subdivision of an inoculant. Consequently, transfer volumes must be removed prior to settling and the dilutions shaken prior to inoculation onto plants.
   (a) Prepare the diluent and dilution blanks using a minimal salt media, such as sterile phosphate–peptone buffer of the following composition: per 1000 mL distilled water: peptone, 1.0 g KH_2PO_4, 0.34 g K_2HPO_4, 1.21 g pH = 7.0. Phosphate–peptone can be used instead of sterile deionized water in order to eliminate a gradient in osmotic potential across the dilution series. Mix the inoculant thoroughly within its bag. Aseptically remove the

Table 1
Plant nutrient solution composition

Component[a]	Final concentration (mg/L PN)	Stock solutions (g/L)	mL/L of stocks to make plant nutrient solutions
Micronutrients			
$CoCl_2 \cdot 6H_2O$	0.004	0.004	
H_3BO_3	2.86	2.86	
$MnCl_2 \cdot 4H_2O$	1.81	1.81	
$ZnSO_4 \cdot 7H_2O$	0.22	0.22	1.0
$CuSO_4 \cdot 5H_2O$	0.08	0.08	
$Na_2MoO_4 \cdot 2H_2O$	0.121	0.121	
$H_2MoO_4 \cdot H_2O$	0.09	0.09	
Macronutrients			
$MgSO_4 \cdot 7H_2O$	492.96	246.48	2.0
K_2HPO_4	174.18	174.18	1.0
KH_2PO_4	136.09	136.09	1.0
$CaCl_2$	110.99	110.99	1.0
$FeC_6H_5O_7 \cdot H_2O$	5.00	5.00	1.0
For complete nutrient solution add 1 mL of NH_4NO_3 (stork = 8 g/100 mL). Add 15 mL of nutrient solution to split pouches and 30 mL to whole pouches			
Stock solution	**Chemical**	**Quantity (g/L)**	**Quantity of stock per liter H_2O**
Without N			
1	K_2SO_4	93	3 mL
2	$MgSO_4 \cdot 7H_2O$	498	1 mL
3	KH_2PO_4	23	1 mL
	K_2HPO_4	145	
4	$CaCl_2$	56	1 mL
5	$CaSO_4$		1 g
6[b]	$FeCl_3$	6.5	1 mL
	Na_2H_2EDTA	13	
7	H_3BO_3	0.23	1 mL
	$MnSO_4 \cdot H_2O$	0.16	
	$ZnSO_4 \cdot 7H_2O$	0.22	
	$CuSO_4 \cdot 5H_2O$	0.08	
	$Na_2MoO_4 \cdot 2H_2O$	0.025	

(continued)

Table 1
(continued)

	$CoCl_2 \cdot 6H_2O$	0.084	
	$NiCl_2$	0.022	
Plus N			
8[c]	KNO_3	10	1 mL
	$(NH_4)_2SO_4$	133	

[a]Weaver and Graham [2]

[b]Dissolve the two chemicals separately before combining

[c]For nutrient solution containing N, substitute stock solution 8 for 1

Table 2
Common method for sterilization and scarification of legume seeds

Treatment	Concentration	Procedure
Surface sterilization		
No treatment	n.a.	Used only after planting many seeds (>500) in *Rhizobium-free* growth containers without subsequent nodulation. Used with commercial quantities of extremely small-seeded, uniformly germinating species that are adversely affected by chemical treatment (e.g., *Trifolium repens*).
Sodium hypochlorite	1–3%	Soak seeds for 1–5 min followed by several (5 or more) rinses with sterile water. Leave seeds in final rinse until imbibed (1–4 h). Useful for all legumes, reduce concentration to treat smaller-seeded species. This treatment should be preceded by a 30 s rinse with 70% ethanol or isopropyl alcohol.
Mercuric chloride	0.2%	Soak seeds for 3 min followed by repeated rinses. Mercuric chloride is toxic and there are difficulties associated with its safe disposal.
Scarification		
Sulfuric acid	conc. (95%)	Immerse dry seeds for 5–60 min depending on the thickness of the seed coat. Drain sulfuric acid, rinse repeatedly (8 or more rinses), leave seeds in final rinse until imbibed (4–24 h). Do not attempt to rinse the seeds by adding water without having first drained as much of the acid away as possible.
Mechanical	n.a.	Nick or abrade seed coat of recently surface sterilized seeds with a sharp, *Rhizobium*-free instrument. Regularly flame sterilize the instrument between samples or groups of samples. Place scarified seeds in sterile water until imbibed.
Hot water	n.a.	Pour boiling water over seeds (1 L/100 g seeds) and allow to stand until cool. If imbibition does not begin, drain and repeat procedure, allow seeds to fully imbibe. This procedure must be preceded by surface sterilization.

inoculant (e.g., with a flamed and cooled spatula), and weigh out 10 g if using 90 mL dilution blanks. Suspend the inoculant in sterile diluent. Place the bottle on a wrist action shaker for 10 min of vigorous mixing. Carry out the tenfold dilution series to the required level.

(b) Serially dilute the final tenfold dilution by pipetting 1–5 mL (transfer volume) into sterile diluent to provide six serial fivefold dilutions. The appropriate amount of diluent can be calculated as follows: diluent volume = [(dilution ratio − 1) × transfer volume] For example, for a dilution ratio of 5 and a transfer volume of 2 mL, the diluent volume is (5 − 1) × 2 = 8 mL. Only a single dilution series per sample need be prepared, as all replicate plant growth units at a given dilution level will be inoculated from the same tube.

(c) Inoculate replicate host plants by pipetting the inoculation volume (usually 1 mL) onto the plant root systems (or into the paper wick trough of growth pouches). Work from the highest dilution level to the lowest dilution level. Plant inoculation is facilitated through the use of a repeating pipette. Those with removable, autoclavable cartridges that dispense adjustable volumes are particularly useful. A single cartridge can be used for each sample by inoculation of highest to lowest dilutions.

6. **Maintenance and watering of growth units:** Given an overall uniformity in growth room environment and the preselection of plants for uniformity prior to inoculation, it is not necessary to randomize plant growth units prior to inoculation. During inoculation, replicate plants should be adjacent to one another to reduce the likelihood of errors in inoculation. The dilution treatments within an MPN test should be systematically arranged. Plants growing in agar slant tubes require little additional maintenance. Growth pouches require periodic watering. Care must be taken not to induce salt stress through evapotranspiration of the nutrient solution. In most cases, addition of sterile water to replace the depleted nutrient solution results in adequate plant growth and development. A common source of nodulated negative controls in an MPN test is the water used for plant maintenance. In all cases, the formation of nodules on uninoculated control plants nullifies the entire MPN test.

3 Results and Analysis

1. **Data collection:** Nodulation occurs 2–4 weeks following inoculation. Nodulation is slowest at higher dilutions. With some

legumes, nodules are still appearing in the high dilutions during the fourth week. Inspect plant root systems for the formation of root nodules 21–28 days following inoculation and record the result for each unit at each dilution. Results are recorded as either positive or negative for each growth unit.

2. **Evaluation of acceptability of results:** The results of a typical assay are presented in Table 3 below which shows a six-step, tenfold serial dilution series with four replicate plants per dilution and an inoculation volume of 1 mL. The number of dilution steps from and including the first not entirely positive to the last not entirely negative dilution is called the Range of Transition (ROT). In the above example, the experimental code is 4-4-3-1-0-0, and the ROT = 2.

   The ROT is a direct measure of experimental compliance with the principle assumptions underlying the MPN procedure, namely, that a single cell is capable of producing a root nodule and that the cells are randomly distributed. Lower values (ROT = 0–2) indicate acceptable results. Higher ROT values (>4) indicate serious problems in experimental technique. The probability of meeting the primary assumptions underlying the MPN technique can be determined for a given ROT, dilution ratio, and number of replicates (Table 4). When the column for tenfold dilutions with four replicates is located on the table, a ROT value of 2 yields a probability of 0.271, and the dilution series is acceptable. An experimental code of 4-2-3-0-2-0 developed under similar experimental conditions has a ROT of 4 and a probability of 0.004. We are certain to a probability value of 0.996 that the results of this dilution series

Table 3
Examples of results of MPN assay

	Replicate				
Dilution level	**1**	**2**	**3**	**4**	**Total positive**
10^{-6}	+	+	+	+	4
10^{-7}	+	+	+	+	4
10^{-8}	+	+	+		3
10^{-9}		+			1
10^{-10}					0
10^{-11}					0

The number of dilution steps from and including the first not entirely positive to the last not entirely negative dilution is called the Range of Transition (ROT). In the above example, the experimental code is 4-4-3-1-0-0, and the ROT = 2

Table 4
Expected frequencies of equalling or exceeding the ROT

	Probability of observing a given ROT value			
	Dilution ratio			
Range of transition	2	4	5	10
Two replicates per dilution level				
1	0.930	0.717	0.660	0.525
2	0.820	0.373	0.281	0.114
3	0.625	0.123	0.075	0.013
4	0.415	0.034	0.015	0.001
5	0.246	0.009	0.003	0.0001
6	0.136	0.002	0.0006	0.00001
Three replicates per dilution level				
1	n.a.	0.891	0.851	0.731
2	n.a.	0.511	0.435	0.193
3	n.a.	0.208	0.123	0.023
4	n.a.	0.060	0.027	0.002
5	n.a.	0.015	0.005	0.0002
6	n.a.	0.004	0.001	0.00002
Four replicates per dilution level				
1	n.a.	0.955	0.931	0.838
2	n.a.	0.682	0.561	0.271
3	n.a.	0.294	0.178	0.035
4	n.a.	0.088	0.040	0.004
5	n.a.	0.023	0.008	0.0004
6	n.a.	0.006	0.002	0.00004
Five replicates per dilution level				
1	n.a.	0.981	0.967	0.899
2	n.a.	0.777	0.661	0.310
3	n.a.	0.371	0.232	0.047
4	n.a.	0.118	0.054	0.005
5	n.a.	0.031	0.011	0.0005
6	n.a.	0.008	0.002	0.00005

Frequency distributions for threefold dilution series are not available
n.a. = frequency distribution not available

Table 5
MPN numbers of two-, three-, four-, five-, and tenfold serial dilutions with two replications

No. of positive results at Dilution level 1-2-3-4-5-6	Population estimate Ratio of the dilution series 3	4	5	10
0-1-0-0-0-0	0.51.0	1.5	4.6	
1-0-0-0-0-0	0.61.2	1.8	2.5	6.0
1-0-1-0-0-0	1.22.5	3.9	5.2	12.0
1-1-0-0-0-0	1.32.6	4.1	5.5	12.0
2-0-0-0-0-0	1.43.3	5.5	8.0	23.0
2-0-1-0-0-0	2.35.4	9.4	14.0	49.0
2-1-0-0-0-0	2.46.0	11.0	16.0	61.0
2-1-1-0-0-0	3.69.3	17.0	28.0	127.0
2-2-0-0-0-0	3.911.0	23.0	40.0	230.0
2-2-0-1-0-0	5.416.7	38.0	71.0	493.0
2-2-1-0-0-0	5.718.4	43.0	81.0	614.0
2-2-1-1-0-0	7.828.1	70.0	142.0	1270.0
2-2-2-0-0-0	8.433.5	91.0	202.0	2305.0
2-2-2-0-1-0	11.451.1	152.0	354.0	4844.0
2-2-2-1-0-0	12.056.1	172.0	408.0	5938.0
2-2-2-1-1-0	1686.4	285.0	713.0	12,500.0
2-2-2-2-0-0	18,104.0	372.0	1016.0	25,000.0
2-2-2-2-0-1	24,162.0	638.0	1797.0	40,000.0
2-2-2-2-1-0	26,179.0	725.0	2109.0	60,000.0
2-2-2-2-1-1	38,285.0	1200.0	3750.0	120,000.0
2-2-2-2-2-0	43,359.0	1600.0	6250.0	120,000.0
2-2-2-2-2-1	72,703.0	2200.0	15,000.0	240,000.0
Confidence factor[b]	2.673.45	4.01	4.47	6.61

[a]This is the population density in the test substrate assuming a 1 mL inoculation volume. Table generated using MPNES software [3]

[b]The population estimate is divided and multiplied by the confidence factor to establish the lower and upper confidence limits ($p = 0.05$), respectively

do not comply with the underlying assumptions and the results are therefore questionable.

3. **Determination of the viable rhizobial population estimate and the 95% confidence levels for the estimate:** Population

Table 6
MPN numbers of two-, three-, four-, five-, and tenfold serial dilutions with three replications

No. of positive results at Dilution level 1-2-3-4-5-6	Population estimate				
	Ratio of the dilution series				
	2	3	4	5	10
1-0-0-0-0-0	0.3	0.7	1.1	1.5	3.5
1-0-1-0-0-0	0.7	1.5	2.3	3.1	7.2
1-1-0-0-0-0	0.7	1.5	2.4	3.2	7.3
2-0-0-0-0-0	0.8	1.7	2.7	3.8	9.1
2-1-0-0-0-0	1.3	28	4.4	6.1	14.0
3-0-0-0-0-0	1.4	3.3	5.5	8.0	23.0
3-0-1-0-0-0	1.9	4.7	8.0	12.0	38.0
3-1-0-0-0-0	2.0	5.0	8.6	13.0	42.0
3-1-1-0-0-0	2.7	6.7	12.0	18.0	74.0
3-2-0-1-0-0	2.8	7.2	13.0	21.0	92.0
3-2-1-0-0-0	3.7	9.8	19.0	31.0	147.0
3-3-0-0-0-0	3.9	11.0	23.0	40.0	230.0
3-3-0-1-0-0	4.8	15.0	32.0	59.0	382.0
3-3-1-0-0-0	5.0	15.0	34.0	64.0	425.0
3-3-1-1-0-0	6.1	20.0	48.0	93.0	738.0
3-3-2-0-0-0	6.4	22.0	53.0	105.0	919.0
3-3-2-1-0-0	8.0	30.0	75.0	156.0	1466.0
3-3-3-0-0-0	8.4	33.0	91.0	203.0	2305.0
3-3-3-0-1-0	10.0	44.0	129.0	298.0	3829.0
3-3-3-1-0-0	11.0	47.0	139.0	322.0	4219.0
3-3-3-1-1-0	13.0	63.0	194.0	467.0	7188.0
3-8-3-2-0-0	14.0	68.0	216.0	532.0	9375.0
3-3-3-2-1-0	17.0	91.0	305.0	791.0	14,375.0
3-3-3-3-0-0	18.0	104.0	372.0	1035.0	25,000.0
3-3-3-3-0-1	22.0	140.0	531.0	1523.0	40,000.0
3-3-3-3-1-0	23.0	149.0	575.0	1680.0	40,000.0
3-3-3-3-1-1	29.0	200.0	800.0	24,220	80,000.0
3-3-3-3-2-0	31.0	220.0	900.0	2969.0	80,000.0
3-3-3-3-2-1	40.0	306.0	1300.0	4219.0	120,000.0

(continued)

Table 6
(continued)

No. of positive results at Dilution level 1-2-3-4-5-6	Population estimate Ratio of the dilution series				
	2	3	4	5	10
3-3-3-3-2-2	53.0	422.0	1900.0	6250.0	160,000.0
3-3-3-3-3-0	44.0	359.0	1700.0	6563.0	120,000.0
3-3-3-3-3-1	60.0	548.0	2600.0	10,000.0	240,000.0
3-3-3-3-3-2	93.0	956.0	5600.0	20,000.0	240,000.0
Confidence factor[b]	2.23	2.75	3.11	3.40	4.67

[a]This is the population density in the test substrate assuming a 1 mL inoculation volume. Table generated using MPNES software [3]

[b]The population estimate is divided and multiplied by the confidence factor to establish the lower and upper confidence limits ($p = 0.0.5$), respectively

estimates are assigned by locating the experimental results on the appropriate MPN table (Tables 5, 6, 7, and 8). These tables are organized by replicate number. Likely results of the dilution series are listed in the first column. When the correct code is located, the researcher obtains a population estimate from the adjacent column which corresponds to the base ratio of the serial dilution (2, 3, 4, 5, or 10).

For example, given the six-step, tenfold dilution series with four replicates which yielded the experimental results 4-4-3-1-0-0 (Table 3) the researcher first locates the table for four replicates (Table 7). The experimental code is located near the center of the experimental results column. The population estimate obtained from the column for tenfold dilutions is 1592 cells/mL. This population estimate reflects the number of cells in the 10^{-5} dilution suspension and assumes a plant inoculation volume of 1.0 mL. This number multiplied by the reciprocal of the level of dilution prior to plant inoculation (10^5) provides the estimate for the population in the inoculant sample (1.6×10^8 viable rhizobia per g inoculant). If inoculation volumes were other than 1.0 mL, additional calculations are required.

The lower 95% confidence limit is obtained by dividing the population estimate by the confidence factor. The upper 95% limit is obtained by multiplying the population estimate by the confidence factor. Confidence factors are listed at the bottom of each population estimate column or can be obtained from Table 9. For the tenfold dilution series with four replicates (Table 7), the confidence factor is 3.80 ($p = 0.05$). The experimental result 4-4-3-1-0-0 resulted in a population estimate of

Table 7
MPN numbers of two-, three-, four-, five-, tenfold serial dilutions with four replications

No. of positive results at	Population estimate				
Dilution level	Ratio of the dilution series				
1-2-3-4-5-6	2	3	4	5	10
1-0-0-0-0-0	0.2	0.5	0.8	1.1	2.5
1-1-0-0-0-0	0.5	1.1	1.7	2.2	5.1
2-0-0-0-0-0	0.5	1.2	1.8	2.5	5.9
2-1-0-0-0-0	0.9	1.9	2.9	3.9	9.2
3-0-0-0-0-0	1.0	2.0	3.3	4.5	11
3-1-0-0-0-0	1.3	2.9	4.6	6.4	15
3-2-0-0-0-0	1.7	3.9	6.2	8.7	21
4-0-0-0-0-0	1.4	3.3	5.5	8.0	23
4-1-0-0-0-0	1.8	4.5	7.7	11	35
4-1-1-0-0-0	2.3	5.7	9.0	15	54
4-2-0-0-0-0	2.4	6.0	10	16	61
4-2-1-0-0-0	2.9	7.5	13	21	92
4-3-0-0-0-0	3.0	8.0	14	24	112
4-3-1-0-0-0	3.7	10	19	32	159
4-3-2-0-0-0	4.4	12	25	43	213
4-4-0-0-0-0	3.8	11	22	40	230
4-4-1-0-0-0	4.7	14	31	57	359
4-4-1-1-0-0	5.5	17	39	75	544
4-4-2-0-0-0	5.6	18	42	81	613
4-4-2-1-0-0	6.6	22	54	107	926
4-4-3-0-0-0	6.8	24	60	121	1123
4-4-3-1-0-0	8.0	30	78	161	1592
4-4-3-2-0-0	9.5	38	102	218	2129
4-4-4-0-0-0	8.4	33	91	202	2305
4-4-4-1-0-0	10	43	125	287	3594
4-4-4-1-1-0	11	53	161	379	5469
4-4-4-2-0-0	12	56	172	409	6137
4-4-4-2-1-0	14	69	221	544	9262
4-4-4-3-0-0	14	74	243	610	11,239

(continued)

Table 7
(continued)

No. of positive results at Dilution level 1-2-3-4-5-6	Population estimate Ratio of the dilution series				
	2	3	4	5	10
4-4-4-3-1-0	17	91	318	830	15,926
4-4-4-3-2-0	20	118	417	1113	21,297
4-4-4-4-0-0	18	104	373	1035	230,545
4-4-4-4-1-0	21	135	518	1464	359,439
4-4-4-4-1-1	25	169	668	1992	546,920
4-4-4-4-2-0	26	179	718	2109	613,730
4-4-4-4-2-1	31	225	681	2812	1,123,930
4-4-4-4-3-0	33	244	876	3281	1,592,630
4-4-4-4-3-1	40	312	961	4375	2,129,690
4-4-4-4-3-2	50	404	1255	5937	3,594,390
4-4-4-4-4-0	43	355	1632	5064	5,469,200
4-4-4-4-4-1	55	485	1650	7177	6,137,300
4-4-4-4-4-2	73	689	2350	10,228	9,262,000
4-4-4-4-4-3	107	1068	3600	15,263	11,239,300
Confidence factor	2.00	2.40	2.67	2.88	3.80

1592 cells, the lower confidence limit ($p = 0.05$) equals 1592/3.80, or 419 cells/mL. The upper confidence limit equals 1592 × 3.80 or 6050 cells. The final results may be expressed as 1.6×10^8 cells per g inoculant (4.2×10^7–6.0×10^8, $p = 0.05$).

4. **Determining population estimates and 95% confidence limits:**
   (a) Locate the appropriate MPN table based on the base dilution ratio of plant applied inoculum and the number of replicates.
   (b) On this table, find the experimental outcome (MPN code) identical to the results obtained from nodulation data. Not all possible experimental outcomes are listed on MPN probability tables. Subjective interpolation is often made, but should be related reasonably to the ROT. To the right of the MPN code is the population estimate corresponding to the result for that particular dilution ratio and number of replicates.

Table 8
MPN numbers of two-, three-, four-, five-, tenfold serial dilutions with five replications

No. of positive results at Dilution level 1-2-3-4-5-6	Population estimate: Ratio of the dilution series				
	2	3	4	5	10
1-0-0-0-0-0	0.2	0.4	0.6	0.8	1.9
1-1-0-0-0-0	0.4	0.8	1.3	1.7	4.0
2-0-0-0-0-0	0.4	0.9	1.4	1.9	4.4
2-1-0-0-0-0	0.7	1.4	2.2	2.9	6.8
3-0-0-0-0-0	0.7	1.5	2.3	3.2	7.7
3-1-0-0-0-0	1.0	2.1	3.3	4.5	10
3-2-0-0-0-0	1.3	2.7	4.3	5.8	13
4-0-0-0-0-0	1.0	2.2	3.6	5.1	12
4-1-0-0-0-0	1.3	3.0	4.8	6.7	16
4-2-0-0-0-0	1.7	3.8	6.1	8.5	21
4-3-0-0-0-0	2.0	4.7	7.6	10	27
5-0-0-0-0-0	1.4	3.3	5.5	8.0	23
5-0-1-0-0-0	1.7	4.1	6.9	10	31
5-1-0-0-0-0	1.7	4.2	7.2	10	32
5-1-1-0-0-0	2.1	5.1	8.9	13	45
5-2-0-0-0-0	2.2	5.3	9.3	14	48
5-2-1-0-0-0	2.6	6.4	11	17	69
5-3-0-0-0-0	2.6	6.7	12	18	78
5-3-1-0-0-0	3.1	8.0	14	23	106
5-3-2-0-0-0	3.6	9.5	18	29	138
5-4-0-0-0-0	3.2	8.5	16	26	127
5-4-1-0-0-0	3.7	10	20	34	168
5-4-2-0-0-0	4.3	12	24	43	215
5-4-3-0-0-0	5.0	14	30	53	270
5-5-0-0-0-0	3.8	11	22	40	230
5-5-0-1-0-0	4.4	13	28	51	312
5-5-1-0-0-0	4.5	13	29	53	327
5-5-1-1-0-0	5.1	15	35	66	452
5-5-2-0-0-0	5.2	16	37	70	488

(continued)

Table 8 (continued)

No. of positive results at Dilution level 1-2-3-4-5-6	Population estimate				
	Ratio of the dilution series				
	2	3	4	5	10
5-5-2-1-0-0	5.9	19	45	88	691
5-5-3-0-0-0	6.1	20	48	94	779
5-5-3-1-0-0	6.9	24	59	119	1070
5-5-3-2-0-0	7.8	28	72	149	1382
5-5-4-0-0-0	7.1	25	64	132	1275
5-5-4-1-0-0	8.1	30	80	170	1690
5-5-4-2-0-0	9.2	37	100	215	2158
5-5-4-3-0-0	10	44	123	268	2715
5-5-5-0-0-0	8.4	33	91	202	2305
5-5-5-0-1-0	9.4	39	113	256	3125
5-5-5-1-0-0	9.6	40	117	268	3281
5-5-5-1-1-0	10	48	141	335	4531
5-5-5-2-0-0	11	50	151	352	4922
5-5-5-2-1-0	12	60	185	442	6875
5-5-5-3-0-0	12	62	196	476	7812
5-5-5-3-1-0	14	75	241	600	10,625
5-5-5-3-2-0	16	89	296	757	13,750
5-5-5-4-0-0	15	79	262	669	13,125
5-5-5-4-1-0	17	95	328	859	16,875
5-5-5-4-2-0	20	115	409	1089	21,250
5-5-5-4-3-0	23	140	509	1357	27,500
5-5-5-5-0-0	18	104	373	1025	22,500
5-5-5-5-0-1	20	124	465	1308	35,000
5-5-5-5-1-0	21	128	484	1357	35,000
5-5-5-5-1-1	24	153	596	1699	60,000
5-5-5-5-2-0	24	160	625	1796	60,000
5-5-5-5-2-1	28	192	768	2304	70,000
5-5-5-5-3-0	29	202	778	2460	80,000
5-5-5-5-3-1	33	244	954	3125	120,000

(continued)

Table 8 (continued)

No. of positive results at Dilution level 1-2-3-4-5-6	Population estimate				
	Ratio of the dilution series				
	2	3	4	5	10
5-5-5-5-3-2	39	295	1167	3906	160,000
5-5-5-5-4-0	35	262	1036	3593	120,000
5-5-5-5-4-1	41	321	1290	4687	160,000
5-5-5-5-4-2	49	395	1602	5937	240,000
5-5-5-5-4-3	59	492	1980	7187	240,000
5-5-5-5-4-4	72	618	2436	9375	240,000
5-5-5-5-5-0	43	356	1463	5781	240,000
5-5-5-5-5-1	52	453	1880	7500	240,000
5-5-5-5-5-2	65	597	2421	10,000	240,000
5-5-5-5-5-3	83	815	3146	17,500	480,000
5-5-5-5-5-4	118	1237	4202	16,728	1,312,535
Confidence factor[b]					

[a]This is the population density in the test substrate assuming a 1 mL inoculation volume. Table generated using MPNES software [3]

[b]The population estimate divided and multiplied by the confidence factor to establish the lower and upper confidence limits ($p = 0.05$), respectively

(c) These tables assume an inoculation volume of 1.0 mL. If a different inoculation volume was applied to the host legumes, the population estimate must be corrected by dividing the inoculation volume into the tabular population estimate.

(d) To obtain the upper and lower confidence limits ($p = 0.05$) multiply and divide the population estimate by the confidence factor found at the base of each MPN table, or obtained from Table 9.

(e) Two population estimates are significantly different from one another when their confidence intervals do not overlap.

5. **Most-Probable-Number Enumeration System (MPNES)—a computer program to calculate population estimates and confidence intervals.** An IBM compatible software program has been developed at NifTAL [3] which provides the population estimate and confidence limits for results of six-step dilution series MPN tests. Users provide the serial dilution ratio

Table 9
Factors for calculating the confidence intervals of Most Probable Number estimates

	Factor for 95% confidence interval at various dilution ratios[a]				
Replicates per dilution	2	3	4	5	10
1	4.01	5.75	7.14	8.31	14.45
2	2.67	3.45	4.01	4.47	6.61
3	2.23	2.75	3.11	3.40	4.67
4	2.00	2.40	2.67	2.88	3.80
5	1.86	2.19	2.41	2.58	3.30
6	1.76	2.04	2.23	2.37	2.98
7	1.69	1.94	2.10	2.23	2.74
8	1.63	1.86	2.00	2.11	2.57
9	1.59	1.79	1.93	2.03	2.44
10	1.55	1.74	1.86	1.95	2.33

[a]Population estimates are multiplied and divided by the confidence factors to establish the upper and lower confidence intervals at ($p = 0.05$), respectively. Confidence factors were calculated using MPNES software [3] after Cochran [4]

and replicate number, and have the option to input the inoculant volume (default = 1.0 mL). The program generates the population estimate and confidence limits for the test sample. MPNES is useful in that solutions can be obtained for experimental results that do not appear in standard MPN tables. MPNES also generates MPN tables for dilution ratios <15 (including non-whole numbers) for 2, 3, 4, 5, and 10 replicates.

References

1. Olsen PE, Sande ES, Keyser HH (1996) Enumeration of rhizobia in inoculant using the most probable number (MPN) plant -- infection assay. In: The enumeration and identification of rhizobial bacteria in legume inoculant quality control procedures. NifTAL Center, Paia, HI, pp 29–56
2. Weaver RW, Graham PH (1994) Legume nodule symbionts. In: Weaver RW, Angle S, Bottomley P, Bezdicek D, Smith S, Tabatabai A, Wollum A, Mickelson SH, Bigham JM (eds) Methods of soil analysis. Part 2 microbiological and biolchemical properties. Soil science society of America, Inc., Madison, pp 200–222
3. Woomer P, Bennett J, Yost R (1990) Overcoming the inflexibility of most-probable-number procedures. Agron J 82:349–353
4. Cochran WG (1950) The comparison of percentages in matched samples. Biometrika 37:256–266

Check for updates

Chapter 3

Isolation and Enumeration of *Azospirillum* by Most-Probable-Number (MPN) Method

Abstract

The genus *Azospirillum* includes the bacteria with plant-growth-promoting traits (PGPB), primarily atmospheric nitrogen fixation in cereal crops like rice through associative symbiosis. It can also fix nitrogen and replace N-fertilizers when associated with grasses, including sugarcane (*Saccharum* spp.), grain crops such as maize (*Zea mays* L.), wheat (*Triticum aestivum* L.), and rice (*Oryza sativa* L.), and pastures such as Brachiaria (*Uruchloa*). Beneficial results have been obtained consistently with *Azospirillum* applied to a variety of crops in dozens of commercial inoculants worldwide. Plant beneficial trait of Azospirillum also relies on the synthesis of phytohormones and other compounds, including auxins, cytokinins, gibberellins, abscisic acid, ethylene, and salicylic acid. Some *Azospirillum* strains can also solubilize inorganic phosphorus, making it more readily available to the plants and resulting in higher yields. Another important feature of Azospirillum is related to biological control of plant pathogens, enabled by the synthesis of siderophores, and limiting the availability of iron (Fe) to phytopathogens, or causing alterations in the metabolism of the host plant, including the synthesis of a variety of secondary metabolites that increase plant resistance to infection by pathogens, a mechanism known as induction of systemic resistance (ISR). Due to the several mechanisms reported to promote plant growth, "theory of multiple mechanisms" is proposed in which the bacterium acts in a cumulative or sequential pattern of effects, resulting from mechanisms occurring simultaneously or consecutively. Population of Azospirillum in the soil is determined by most-probable-number (MPN) technique.

Keywords *Azospirillum*, *Enumeration*, Most-probable-number (MPN), *Nitrogenase*

1 Principle

The MPN method permits estimation of population density without an actual count of single cells or colonies. It is sometimes called the method of ultimate or extinction dilution or end point dilution. A prerequisite of this method is that microorganisms whose population is to be determined must be able to bring some characteristics and readily recognizable transformation in the medium into which it is inoculated or else the microorganisms itself, after it has undergone multiplication, must be easily recognizable in the substrate. *Azospirillum* is a micro-aerophilic nitrogen fixer. A semisolid N-free malate

M. Senthilkumar, N. Amaresan and A. Sankaranarayanan, *Plant-Microbe Interactions: Laboratory Techniques*, Springer Protocols Handbooks, https://doi.org/10.1007/978-1-0716-1080-0_3,

medium with bromothymol blue dye is used for enumeration. The presence of *Azospirillum* in the tube is indicated by (a) a change in dye colour from green to blue, (b) formation of pellicle at the subsurface of medium, and (c) reduction of acetylene into ethylene.

2 Procedure

1. Prepare glass tubes (25) each containing 5 mL of semisolid malate medium. Sterilize them at 121 °C, 15 lb pressure.
2. Take 1 g soil and prepare tenfold dilution series up to 10^{-6}.
3. Withdraw by sterile pipette 1 mL aliquots from dilutions 10^{-2}, 10^{-3}, 10^{-4}, 10^{-5}, and 10^{-6}.
4. For each dilution five tubes have to be inoculated with 1 mL aliquot separately.
5. Incubate the inoculated tubes for 2 days at 30 °C.
6. Examine each tube for color change and pellicle formation. Record the growth in each individual tube as either positive or negative. Visual observation can be confirmed by estimation of nitrogenase activity of each tube. The tubes positive for Azospirillum should show nitrogenase activity.
7. Calculate the number of positive tubes in the least concentrated dilution in which the greater number of tubes is positive and designate as P1. The numbers of positive tubes in the next two higher dilutions are considered as P2 and P3.
8. Use the table of MPN for use with tenfold dilution and five tubes per dilution [1]. The find the row of numbers in MPN table in which P1 and P2 correspond to the values observed experimentally. Follow the row numbers across the table to the column headed by observed value of P3. The figure at the point of intersection is the most probable number of organisms in the quantity of original sample represented in the inoculums added in the second dilution (P2). Multiply this figure with appropriate dilution factor to obtain the MPN for the original sample.

   Example: Suppose the following observations are made:

Dilution	+ve tubes
10^{-3}	5
10^{-4}	3
10^{-5}	1
10^{-6}	0

In this series, P1 = 5, P2 = 3, and P3 = 1, For this combination of P1, P2, and P3, the MPN table gives 1.1 as the most probable number of organisms in the quantity of inoculums applied in the 10^{-5} (P2) dilution. Multiplying this number with dilution factor 10^5, gives 1.1×10^5 as the MPN for original sample.

Reference

1. Cochran WG (1950) Estimation of bacterial densities by means of the Most Probable Number. Biometrics 6:105–116

Check for updates

Chapter 4

Quantitative Estimation of Leghemoglobin Content in Legume Root Nodules

Abstract

The oxygen-binding hemoprotein was first recognized in root nodules by Kubo in 1939. Leghemoglobin (Lb) occurs in the infected cells of legume root nodules. The concentration ranges 1–3 mg per g fresh weight depending on the plant species. Heme synthesis for Lb seems to be a bacterial property whereas, the apoprotein (globin) is coded for by the plant DNA and thus a well-evolved coexistence mechanism between the host and the bacterium is seen in legume root nodules. Lb facilitates oxygen diffusion across the nodule into the nitrogen-fixing bacteroids to support oxidation and at the same time ensures the oxygen sensitive nitrogenase activity without damage. This protocol deals with the determination of leghemoglobin content in legume nodule through spectrophotometer.

Keywords Leghemoglobin, Oxygen diffusion, Root nodule, *Rhizobium*, Symbiotic efficiency

The oxygen-binding hemoprotein was first recognized in root nodules by Kubo in 1939. Leghemoglobin (Lb) occurs in the infected cells of legume root nodules. The concentration ranges 1–3 mg per g fresh weight depending on the plant species. Heme synthesis for Lb seems to be a bacterial property whereas, the apoprotein (globin) is coded for by the plant DNA and thus a well evolved coexistence mechanism between the host and the bacterium is seen in legume root nodules. Lb facilitates oxygen diffusion across the nodule into the nitrogen-fixing bacteroides to support oxidation and at the same time ensures the oxygen sensitive nitrogenase activity without damage [1].

1 Principle

Hemoglobin reacts with pyridine in strong alkali to produce hemochrome. The hemochrome is measured at 556 nm.

M. Senthilkumar, N. Amaresan and A. Sankaranarayanan, *Plant-Microbe Interactions: Laboratory Techniques*, Springer Protocols Handbooks, https://doi.org/10.1007/978-1-0716-1080-0_4, © Springer Science+Business Media, LLC, part of Springer Nature 2021

2 Materials

1. Diluent Buffer: 0.1 M sodium/potassium phosphate buffer (pH 7.4).
2. Alkaline Pyridine Reagent: Dissolve 0.8 g NaOH in 50 mL water and cool. Add 33.8 mL of pyridine (33.2 g), dissolve and dilute to 100 mL with water. This produces 4.2 M pyridine in 0.2 M NaOH.
3. Sodium Dithionite: Grind finely and store in small stoppered tubes in a desiccator.
4. Potassium Hexacyanoferrate.

3 Method

1. *Extraction:* Mix the fresh or thawed nodules with 1–3 volumes of phosphate buffer and macerate in a mixer. Filter through two layers of cheesecloth. Discard the nodule debris. Clarify the turbid reddish brown filtrate by centrifugation at 10,000 g for 10–30 min. Dilute suitably.
2. To a suitable volume (2–5 mL) of the extract add an equal volume of alkaline pyridine reagent and mix. The solution becomes greenish-yellow due to the formation of ferric hemochrome.
3. Divide the hemochrome equally between two tubes.
4. To one portion add a few crystals of sodium dithionite to reduce the hemochrome. Stir without aeration. Read at 556 nm after 2–5 min against a reagent blank.
5. To the other portion add a few crystals of potassium hexacyanoferrate to oxidize the hemochrome and read at 539 nm.
6. Calculate $A_{556} - A_{539}$.

$$\text{Lb concentration (mM)} = \frac{A_{556} - A_{539} \times 2D}{23.4}$$

Where, D is the initial dilution. (The calculation is based upon the equation $E = 23.4 \times 10^3$ mol/cm.)

4 Notes

1. Best yields of Lb are obtained when the nitrogen-fixing activity is near the maximum. For most legume crops, 30- to 35-day--old nodules are suitable.
2. To test field-grown crops, it is essential that the soil is low in nitrogen and heavily inoculated with Rhizobium.

3. Nodules can be picked either into liquid nitrogen or ice-cold 0.1 M phosphate buffer pH 7.4. Nodules picked in this way can be stored frozen for short periods.

   In purified preparations, Lb can be estimated by measuring the reduced hemochrome. The formula is used for calculation since $E = 33.9 \times 10^3$ mol/cm for reduced hemochrome.

$$\text{Lb concentration (mM)} = \frac{A_{556} \times 2D}{33.9}$$

Reference

1. Appleby CA, Bergersen FJ (1980) In: Bergersen FJ (ed) Methods for evaluating biological nitrogen fixation. John Wiley and Sons, New York, NY, p 315

Check for updates

Chapter 5

Quantitative Estimation of Nitrogenase Activity: Acetylene Reduction Assay

Abstract

Symbiotic nitrogen fixation plays an important role in agriculture, and there has been a goal to improve symbiotic efficiency to reduce the use of chemical fertilizers. Symbiotic nitrogen fixation is catalyzed by the molybdenum nitrogenase enzyme. The molybdenum nitrogenase is composed of MoFe protein (NifDK) and Fe protein (NifH). The MoFe protein is an $\alpha_2\beta_2$ heterotetramer that contains the iron–molybdenum cofactors (FeMo-co) and P clusters. The FeMo-co is a [Mo–7Fe–9S–C-homocitrate] cluster which serves as the active site of nitrogen binding and reduction. The P-cluster is a [8Fe–7S] cluster which shuttles electrons to the FeMo-co. The Fe protein is a γ_2 homodimer bridged by an intersubunit [4Fe–4S] cluster that serves as the obligate electron donor to the MoFe protein. Nitrogenase is a versatile enzyme capable of catalyzing the reduction of several substrates other than nitrogen, including acetylene, azide, nitrous oxide, nitriles, and isonitriles. Ethylene produced due to the reduction of acetylene by nitrogenase is determined by using gas chromatography.

Keywords Nitrogenase, Acetylene, Ethylene, Gas chromatography

1 Principle

Nitrogenase is a versatile enzyme capable of catalyzing the reduction of several substrates other than nitrogen, including acetylene, azide, nitrous oxide, nitriles, and isonitriles [1]. The observation that acetylene is an inhibitor of dinitrogen fixation and converted into ethylene by nitrogenase provided the basis for the estimation of N2 fixation. The C_2H_2 reduction method provides a simple, inexpensive, highly sensitive, and nondestructive procedure for measuring rates of N_2-fixation. It can be used to measure N-fixation in free living N-fixing bacteria, blue green algae, associative symbiosis, and nodule bearing leguminous and nonleguminous plants.

The reduction of N_2 and C_2H_2 is identical and designated as follows.

M. Senthilkumar, N. Amaresan and A. Sankaranarayanan, *Plant-Microbe Interactions: Laboratory Techniques*, Springer Protocols Handbooks, https://doi.org/10.1007/978-1-0716-1080-0_5,

$$N_2 + 6H^+ + 6e^- \rightarrow 2NH_3$$
$$C_2H_2 + 2H^+ + 2e^- \rightarrow C_2H_4$$

Thus the theoretical ratio of the substrate reduced is 3:1. Which is included in the equation used to quantify the rate of nitrogen fixed based on C_2H_2 reduction.

mgN = (mM C_2H_4 × 28)/3

ARA technique is an indirect method of measuring N_2-fixation at a point of time. The technique involves incubation of nitrogenous containing system in a known atmosphere of C_2H_2 (5–20%) in a gas phase, and after an optimum time of incubation, C_2H_4 produced is measured by a gas chromatograph using a flame ionization detector (FID).

2 Determination of Nitrogenase Activity of Free Living Bacteria

1. Prepare slants or use semisolid medium. *Azospirillum*—Semisolid N-free malate medium [2]; *Azotobacter*—Jensen's N-free medium [3, 4]; Ex-planta nitrogenase activity of *Rhizobium*—CS medium [5].
2. Streak loopful of cultures on the slants and incubate at 28 °C for 2–7 days.
3. Replace cotton plugs with air tight serum stoppers.
4. Remove 10% of atmospheric air from the tube and inject equal volume of acetylene using a syringe and incubate at 28 °C for 16–24 h.
5. Draw 1 mL of gas sample from the tube and inject into gas chromatograph for ethylene estimation.
6. Similarly record the values for 1 mL of standard ethylene gas.
7. Estimate the total protein of each tube as follows. (a) Collect bacterial cells in 2 mL of 2N NaOH, (b) keep cell suspension in boiling water bath for 10 min, (c) cool it, neutralize with 2 mL of 2N HCl and estimate the protein by Lowry's methods or Bradford method.
8. Express the C_2H_4 produced in *n* mol of C_2H_4 produced/h/mg protein.

3 Acetylene Reduction Assay of Root Nodule System

3.1 Procedure

1. Uproot legume plants, shake gently to remove adhering soil.
2. Detach the root system along with the nodules.
3. Transfer the roots along with nodules into assay vial.
4. Insert serum stopper and replace 10% air with acetylene and incubate for 1 h.

5. Carry gas chromatograph as described earlier. Note: Determine the volume of airspace in the vial by adding the water to the vial in the presence of root samples and measuring the volume of water.
6. Detach the root nodules from the roots, dry them in hot air oven at 80 °C for 3 days and take dry weight. Express the C_2H_4 produced in *n* mol of C_2H_4 produced/h/mg protein.

References

1. Hardy RWF, Burus RC, Holsten RD (1973) Application of the acetylene ethylene assay for measurements of nitrogen fixation. Soil Biol Biochem 5:47–81
2. Okon Y, Albrecht SL, Burris RH (1977) Methods for growing *Spirillum lipoferum* and for counting it in pure culture and in association with plants. Appl Environ Microbiol 33:85–88
3. Jensen HL (1954) The Azotobacteriaceae. Bacteriol Rev 18:195–213
4. Norris JR, Chapman HM (1968) Classification of *Azotobacter*. In: Gibbs BM, Shapton DA (eds) Identification methods for microbiologists. Academic Press, London; New York, NY, pp 19–27
5. Pagan JD, Child JJ, Scowcroft WR, Gibson AH (1975) Nitrogen fixation by *Rhizobium* cultured on a defined medium. Nature 256:406–407

Check for updates

Chapter 6

Quantitative Estimation of Nitrogen Fixation: $^{15}N_2$ Natural Abundance Study

Abstract

The ^{15}N natural abundance method has been successfully applied to the study of N_2-fixation by legumes. ^{15}N Natural abundance technique is based on the principle that the proportion of N derived from the air (%Ndfa) via biological N_2-fixation (BNF) is proportional to the difference in ^{15}N abundance between the "N_2-fixing" legume and that of a suitable reference plant which only obtains nitrogen from the soil. The ^{15}N natural abundance isotope-based method is the most commonly employed, because it provides a time-integrated measure of the total amount of nitrogen fixed over the entire growth period. Moreover, this ^{15}N-based technique has the potential to separately evaluate the inputs of BNF- and soil-derived nitrogen to the legume based on a single harvest of the legume crop at maturity. The strategy applied to exploit this technique is to assume that reference plants which are known to be unable to obtain nitrogen from N_2-fixation accumulate nitrogen only from the soil. Thus, if N_2-fixing legumes (e.g., such as soybean (*Glycine max*), French/common bean (*Phaseolus vulgaris*), cowpea (*Vigna unguiculata*)) have significantly lower ^{15}N abundance than the reference plants, then the difference can be interpreted quantitatively to assess the contribution of biological nitrogen fixation.

Keywords ^{15}N Natural abundance, Isotope, Nitrogenase, Symbiotic efficiency

The ^{15}N natural abundance method has been successfully applied to the study of N_2-fixation by legumes. ^{15}N Natural abundance technique is based on the principle that the proportion of N derived from the air (%Ndfa) via biological N_2-fixation (BNF) is proportional to the difference in ^{15}N abundance between the "N_2-fixing" legume and that of a suitable reference plant which only obtains nitrogen from the soil. The ^{15}N natural abundance isotope-based method is the most commonly employed, because it provides a time-integrated measure of the total amount of nitrogen fixed over the entire growth period. Moreover, this ^{15}N-based technique has the potential to separately evaluate the inputs of BNF- and soil-derived nitrogen to the legume based on a single harvest of the legume crop at maturity.

M. Senthilkumar, N. Amaresan and A. Sankaranarayanan, *Plant-Microbe Interactions: Laboratory Techniques*, Springer Protocols Handbooks, https://doi.org/10.1007/978-1-0716-1080-0_6, © Springer Science+Business Media, LLC, part of Springer Nature 2021

The strategy applied to exploit this technique is to assume that reference plants which are known to be unable to obtain nitrogen from N_2-fixation accumulate nitrogen only from the soil. Thus, if N_2-fixing legumes (e.g., such as soybean (*Glycine max*), French/common bean (*Phaseolus vulgaris*), cowpea (*Vigna unguiculata*)) have significantly lower ^{15}N abundance than the reference plants, then the difference can be interpreted quantitatively to assess the contribution of biological nitrogen fixation.

1 Procedure

The ^{15}N natural abundance method can be applied both in greenhouse and field experiments. It allows the estimation of N_2-fixation in locations, where legumes and non N-fixing plants (usually non legumes) are present.

1. Selection of reference plant: Choose one or more reference plant (a non N-fixing plant) and include it in the experiment. Collect the samples of reference plants from each experimental treatment and preferably from each replicate plots. In greenhouse trials very often specific reference treatments are included in the experimental design, whereas in farmers' field weeds can also be used as reference plant. The choice of reference plant must be very careful, since the reference plant must have almost the same biological cycle as the test legume crop. Another point relates the choice of the reference plant is that it has to follow the same N pattern as the legume crop. The reference plant must exploit the same N soil as your main legume crop. Reference plant must be sampled at the same growth stage as the legume crop.
2. The calculation of the proportion of N in the legume plant derived from the air (%Ndfa) is calculated using the equation of Shearer and Kohl [1]:

   $\%Ndfa = (\delta 15N(\text{reference}) - \delta 15N(\text{N-fixing})/\delta 15N(\text{reference}) - B) \times 100$

   B value is the δ15N of shoots of legumes that are fully dependent upon N_2-fixation and sampled at the same growth stage as the field plants.
3. To estimate the "*B*" value, establish a greenhouse trial where plant must be grown in an inert substrate like sterilized sand without N fertilization. **The substrate must not contain any N source**. The plants must be inoculated with the same strain that you are testing in the field. The greenhouse trial must provide a reasonable "*B*" value to be used in the field experiment. Otherwise, it can be used "*B*" value from literature can also be used.
4. **Sampling:** Whole shoots should be sampled. Do not split your plant material into (leaves, stems, pods). Legumes and

reference plants must be sampled at the same time before reaching its maturity, preferably at the stage of early pod-filling. If roots from pots can be harvested a whole plant can be determined for 15N. Nodules can also be harvested and analyzed for 15N.

5. **Sample processing:** Divide plant material in to shoots, roots, nodules contained with large brown paper bags. Dry in an oven at 70 °C until the plant material reaches its stable weight. Milled each sample to a fine powder: not very fine (i.e., not like bread flour), but resembling ground black pepper. Nodule samples need ground in a mortar and pestle or RetchTM (ball) mill. Store each ground sample in a sealed (airtight), glass or plastic container.
6. **Weighing for mass spectrometry (Total-N and 15N determinations):** Weigh 0.3–0.4 (MAX) mg of powdered plant material into tin capsules. For *soil samples*: use not more than 5–6 (MAX.) mg per capsule. Record the exact weight of each sample. First crimp the capsule by clasping the top together and fold-over toward the base, reshaping the capsule into a tubular roll. Take one end of the roll and fold again, this time along the length of the role to form a tight coil. Ensure there are no sharp edges. Place each sample for processing into a labeled 96 well microtiter (ELISA) plate, and note the orientation/recording the position (row and column coordinates), and weight of each sample in a sample record sheet. Analyze with a mass spectroscopy following standard procedure.

Reference

1. Shearer G, Kohl DH (1988) Natural ^{15}N Abundance as a method of estimating the contribution of biologically fixed nitrogen to N_2-fixing systems: potential for non-legumes. Plant Soil 110(2):317–327

Check for updates

Chapter 7

Estimation of Glutamine Synthetase (GS), Glutamate Synthase (GOGAT), and Glucose Dehydrogenase (GDH)

Abstract

Plants utilize nitrate, ammonium, and dinitrogen (N_2) molecules as external nitrogen sources. N_2 is reduced to NH_3 during the rhizobium-legume symbiosis. Ammonia a primary N_2 fixation product in bacteroids of a symbiotic form of rhizobia in nodules is rapidly excreted to the plant cytosol and assimilated into glutamine via the GS/GOGAT system. Ammonium is the final form of inorganic nitrogen prior to the synthesis of organic nitrogen compounds. Ammonium is assimilated into the Gln amide group, which is then transferred to the position of 2-oxoglutarate, yielding two molecules of Glu by the concerted reaction of Gln synthetase (GS; EC 6.1.1.3) and Glu synthase (ferredoxin [Fd]-GOGAT; EC 1.4.7.1; NADHGOGAT; EC 1.4.1.14). Nitrogen is then incorporated into Asp, Ala, Asn, and other amides and amino acids. Gln-dependent Asn synthetase (AS; EC 6.3.5.4) provides Asn, which serves as a nitrogen carrier together with Gln and Glu. Ammonium might be directly incorporated into Glu by amination of 2-oxoglutarate via mitochondrial Glu dehydrogenase (NADH-GDH; EC 1.4.1.2) and subsequently into Gln by cytosolic GS1 under particular physiological conditions.

Keywords Gln synthetase (GS), Glutamine oxoglutarate aminotransferase (GOGAT), Glucose dehydrogenase (GDH), Assimilation

Ammonia a primary N_2 fixation product in bacteroids of a symbiotic form of rhizobia in nodules is rapidly excreted to the plant cytosol and assimilated into glutamine via the GS/GOGAT system.

1 Method

1.1 Enzyme Extraction

1. Grind 1 g of leaf sample in 4 mL of grinding medium consists tricine buffer (165 mM), sucrose (0.4 M), potassium chloride (10 mM), magnesium chloride (10 mM), EDTA (10 mM), and β-mercaptoethanol (10 mM) [3].
2. Centrifuge the homogenate for 7 min at 500 g to remove unbroken cells and cell fragments. Again centrifuge the super-

M. Senthilkumar, N. Amaresan and A. Sankaranarayanan, *Plant-Microbe Interactions: Laboratory Techniques*, Springer Protocols Handbooks, https://doi.org/10.1007/978-1-0716-1080-0_7, © Springer Science+Business Media, LLC, part of Springer Nature 2021

natant solution at 10,000 g for 15 min to yield a crude mitochondrial pellet and supernatant.

3. Centrifuge the supernatant solution again at 100,000 g for 30 min to yield a clear solution for GS and GOGAT assays. The high speed centrifugation is necessary to remove NADH oxidase which otherwise interfered with the GOGAT and GDH assay.

2 In Vitro Glutamine Synthetase (GS) Assay [1]

1. Take 0.5 mL of the synthetase assay mixture contains glutamate (50 mM), hydroxylamine hydrochloride (5 mM), magnesium sulphate (50 mM) and ATP (20 mM) in tris-HCl (100 mM, pH 7.8). The pH values of glutamate and hydroxylamine solution should be carefully adjusted to pH 7.8. Prepare fresh ATP solution before each assay and adjust the pH to just above 7.0.
2. Initiate the reaction by the addition of 0.2 mL of enzymes extract and incubate at 30 °C. Terminate the reaction by the addition of 0.7 mL of $FeCl_3 \cdot 6H_2O$ in 0.02N HCl. Incubated tubes with enzyme extract and all the reagents except ATP serve as control. For precipitation of protein, centrifuge the content in tubes at 10,000 g for 5 min. The absorbance of the supernatant was determined at 540 nm. Develop the standard curve using L-glutamic acid γ-monohydroxamate to obtain an accurate quantification of GS activity.

3 In Vitro Glutamate Synthase (GOGAT) Assay [2]

1. Prepare the supernatant/enzyme preparation as described earlier in this protocol.
2. The reaction mixture contained in a final volume of 3 mL tris–HCl buffer (75 μmol, pH 7.5), α-ketoglutarate (10 μmol), L-glutamine (15 μmol), NADH (0.3 μmol) and enzyme preparation (0.5 mL).
3. Initiate the reaction by adding NADH and measure the rate at 340 nm at 25 °C for 5 min. The control lacked glutamine, NADH, and α-ketoglutarate.
4. Calculate the enzyme activity using the extinction coefficient of NADH at 340 nm i.e., 6.3 × 103 L mol/cm.

4 In Vitro Glutamate Dehydrogenase (GDH) Assay [2]

1. Prepare the supernatant/enzyme preparation as described earlier in this protocol.

2. The reaction mixture contains in a final volume of 3 mL phosphate buffer (75 μmol, pH 7.5), α-ketoglutarate (10 μmol), ammonium chloride (300 μmol), NADH (0.3 μmol) and suspension medium containing enzyme (0.2 mL).
3. Initiate the reaction by adding NADH, and the oxidation of NADH is followed at 340 nm and at 25 °C for 5 min. The blank consists of the complete reaction mixture without ammonium chloride.
4. Calculate the enzyme activity using the extinction coefficient of NADH at 340 nm i.e., 6.3 × 103 L mol/cm.

References

1. Lea PJ, Blackwell RD, Chen FL, Hetch U (1990) Enzymes of ammonia assimilation. In: Lea PJ (ed) Methods in plant biochemistry, vol 3. Academic Press, London, pp 257–276
2. Mohanty B, Fletcher JS (1980) Ammonium influence on nitrogen assimilatory enzymes and protein accumulation in suspension cultures of pearl scarlet rose. Physiol Plant 48:453–459
3. Cooper TG, Beevers H (1969) Mitochondria and glyoxysomes from castor bean endosperm. J Biol Chem 244:3507–3513

Chapter 8

Colorimetric Analysis of Ureide-N, Amino-N, and Nitrate-Nitrogen

Abstract

Nodulated soybean plants accumulate ureide compounds, allantoin, and allantoic acids in the shoots. Comparing $^{15}N_2$ fixed in nodules and $^{15}NO_3^-$ absorbed in the roots, it was confirmed that most of the ureides in the shoots derived from nodules. Similar $^{15}N_2$ fixation studies revealed that ammonia a primary N_2 fixation product in bacteroids of a symbiotic form of rhizobia in nodules is rapidly excreted to the plant cytosol and assimilated into glutamine via the GS/GOGAT system, and synthesized to ureide via de novo synthesis of purine base. Many tropical grain legumes, such as soybean, common bean, cowpea, pigeon pea, and mung bean that have spherical determinate type of nodules transport the bulk of fixed N as ureide (allantoin and allantoic acid). On the other hand, nitrate and amino acids (especially amide, asparagine) are the major forms of N derived from soil and fertilizer N transported by roots. Relative ureide method was also developed for the evaluation of % Ndfa by analyzing the nitrogen composition of xylem sap obtained from bleeding sap from a cut stump, or vacuum collection from shoot. The concentration of ureide-N, nitrate-N, and amino-N can be easily determined by colorimetry.

Keywords Ureide-N, Amino-N, Nitrate-N, Allantoin, Allantoic acid

Nodulated soybean plants accumulate ureide compounds, allantoin and allantoic acids in the shoots. Comparing $^{15}N_2$ fixed in nodules and $^{15}NO_3^-$ absorbed in the roots, it was confirmed that most of the ureides in the shoots derived from nodules. Similar $^{15}N_2$-fixation studies revealed that ammonia a primary N_2-fixation product in bacteroides of a symbiotic form of rhizobia in nodules is rapidly excreted to the plant cytosol and assimilated into glutamine via the GS/GOGAT system, and synthesized to ureide via de novo synthesis of purine base [1].

Many tropical grain legumes, such as soybean, common bean, cowpea, pigeon pea, and mung bean that have spherical determinate type of nodules transport the bulk of fixed N as ureide (allantoin and allantoic acid). On the other hand, nitrate and amino acids (especially amide, asparagine) are the major forms of N derived from soil and fertilizer N transported by roots [2, 3].

M. Senthilkumar, N. Amaresan and A. Sankaranarayanan, *Plant-Microbe Interactions: Laboratory Techniques*, Springer Protocols Handbooks, https://doi.org/10.1007/978-1-0716-1080-0_8, © Springer Science+Business Media, LLC, part of Springer Nature 2021

Herridge et al. [4] developed the relative ureide method for evaluation of % Ndfa by analyzing the nitrogen composition of xylem sap obtained from bleeding sap from a cut stump, or vacuum collection from shoot. The concentration of ureide-N, nitrate-N, and amino-N can be easily determined by colorimetry.

This method is reliable in the field experiment of soybean, without any requirement of reference plants. It is the easiest way to measure % Ndfa in farmers' field, because no preparation is necessary before sampling. This method is also applicable for experiments with variable N fertilizer application. In field conditions, the simple equation can be adapted for the estimation of % Ndfa.

$$\%\mathrm{Ndfa} = [\mathrm{Ureide} - \mathrm{N}/(\mathrm{Ureide} - \mathrm{N} + \mathrm{Nitrate} - \mathrm{N} + 2 \times \alpha - \mathrm{Amino} - \mathrm{N})] \times 100$$

Note: The original equation proposed by Herridge was "α-Amino-N" instead of "2 × α-Amino-N" in the above equation. Based on the analysis of root bleeding xylem sap amino acids compositions, 2N amide Asparagine was the major amino acid throughout the stages and the average N number in amino acids was 1.7, hence "2 × α-amino-N" is currently used for the estimation.

1 Determination of N-Fixing Activity and Absorption Rate

1. By periodical sampling of legume plants and xylem sap, quantitative estimation of the seasonal changes in N_2-fixation activity and N absorption rate is possible. Soybean plants can be sampled at four or five times, at R1 (initial flowering), R3 (maximum shoot growth), R5 (pod filling), and R7 (yellow leaf) stages for xylem sap and plant N analyses.
2. Harvest the soybean plants at R1, R3, R5, and R7 stages and consider the same for D1, D2, D3, and D4, respectively. D0 is the day of sowing. Determine the N content of soybean shoot or whole plant including roots by Kjeldahl digestion method. Average daily N increase in plant is calculated between successive sampling. Daily N gain ΔN2–3 = (N3 − N2)/(D3 − D2).
3. Calculate ureides-N value (RU%) at each sampling time. The average RU% is calculated as follows: RU%2–3 = (RU% 2 + RU % 3)/2. Exceptionally, the first phase between D0 and D1, RU % 0–1 = RU%1. Combining the data of daily N gain and average RU%, the daily N2-fixation activity and daily N absorption rate can be calculated as follows:

$$\text{Daily N2} - \text{fixation activity} = \Delta\text{N2} - 3 \times \text{RU\%2} - 3/100/\text{D2} - 3$$

$$\text{Daily N absorption rate} = \Delta N2-3 \times (100 - RU\%2-3)/100/D2-3$$

2 Determination of Ureide-N [5]

1. Cut the stem of healthy plants and collect 50 μL of the bleeding xylem sap into a test tube.
2. Add 3 mL of 0.083 M NaOH solution to the xylem sap.
3. Heat the test tubes in boiling water bath for 8 min. All the allantoin is degraded into allantoic acid by alkali hydrolysis.
4. Cool the test tubes in cool water.
5. Add 1 mL of ice-cooled phenylhydrazinium solution. Phenylhydrazinium solution: Dissolve 0.33 g of phenylhydrazinium chloride in 100 mL of water, and add 100 mL of 0.65 M HCl solution. Cool in ice water bath before use.
6. Heat the test tubes in boiling water bath for 2 min. All the allantoic acid is degraded into urea and glyoxilic acid.
7. Cool the tubes immediately in ice-water bath for 15 min.
8. Add 2.5 mL of ferricyanide solution and wait for 30 min in ice-water bath.
9. Ferricyanide solution: Dissolve 1.67 g of potassium ferricyanide in 100 mL of water. Add 400 mL of 10 M HCl solution. Cool in ice water before use.
10. Measure the absorbance at 520 nm by spectrometry. Standard 5 mM allantoin solution (280 μgN mL): Dissolve 197.7 mg of allantoin in 250 mL of water.

3 Determination of Amino-N by the Ninhydrin Method [6]

1. Collect 50 μL of xylem sap into a test tube.
2. Add 1.5 mL of citrate buffer into the tube. Citrate buffer: Dissolve 56 g of citrate and 21.3 g of NaOH in 1 L of water.
3. Add 1.2 mL of ninhydrin solution. Ninhydrin solution: Dissolve 0.958 g of ninhydrin and 33.4 mg of ascorbic acid in 3.2 mL of water. Then solution is mixed with methoxyethanol (methylcellosolve) up to total of 100 mL.
4. Boil the test tube in boiling water bath for 20 min with aluminum foil lid.
5. Add 3 mL of 60% ethanol (60 mL of ethanol plus 40 mL of water), and cool until room temperature.

6. Measure the absorbance at 570 nm by spectrometry.
7. Standard 10 mM amino acid solution (140 μg α-amino-N mL-1): Dissolve 132 mg of asparagine (150 mg of asparagine monohydrate) plus 146 mg glutamine in 200 mL of water. This solution contains 5 mM asparagines plus 5 mM glutamine. Total N concentration is 280 μg N mL^{-1}. Be careful to use α-amino-N concentration for the calculation of simple relative ureide equation with "2 × α-amino-N."

4 Determination of Nitrate-N [7]

1. Collect 50 μL of xylem sap into a test tube.
2. Add 200 μL of salicylic acid-sulfate solution into the tube, mix well and wait for 20 min. Salicylic acid-sulfate solution: Dissolve 5 g of salicylic acid in 100 mL of concentrated sulfuric acid.
3. Add 5 mL of 2 M NaOH solution into tube. Mix well and wait for 20 min.
4. Measure the absorbance at 410 nm.
5. Standard nitrate solution (70 μg N mL^{-1}): Dissolve 425 mg of sodium nitrate in 1 L of water.

References

1. Ohyama T, Kumazawa K (1978) Incorporation of ^{15}N into various nitrogenous compounds in intact soybean nodules after exposure of $^{15}N_2$ gas. Soil Sci Plant Nutr 24:525–533
2. Ohyama T, Kumazawa K (1979) Assimilation and transport of nitrogenous compounds originated from $^{15}N_2$-fixation and $^{15}NO_3$ absorption. Soil Sci Plant Nutr 25:9–19
3. Ohtake N, Nishiwaki T, Mizukoshi K, Minagawa R, Takahashi Y, Chinushi T, Ohyama T (1995) Amino acid composition in xylem sap of soybean related to the evaluation of N_2-fixation by the relative ureide method. Soil Sci Plant Nutr 41:95–102
4. Herridge DF, Bergersen FJ, Peoples MB (1990) Estimates of N2 fixation by soybean in the field using the ureide and of nitrogen fixation by soybean in the field using the ureide and natural 15N abundance methods. Plant Physiol 93:708–716
5. Young E, Conway CF (1942) On the estimation of allantoin by the Remini-Shryver reaction. J Biol Chem 142:839–853
6. Herridge DF (1984) Effects of nitrate and plant development on the abundance of nitrogenous solutes in root-bleeding and vacuum-extracted exudates of soybean. Crop Sci 25:173–179
7. Cataldo DA, Haroon M, Schrader LE, Youngs VL (1975) Rapid colorimetric determination of nitrate in plant tissue by nitration of salicylic acid. Commun Soil Sci Plant Anal 6:71–80

Part II

Techniques for Studying Mineral Solubilization and Mobilization

Chapter 9

Isolation of Mineral Phosphate Solubilizing Microbes from Rhizosphere Soil

Abstract

Phosphorus (P) is the second most important key element for plant growth in terms of quantitative requirement after nitrogen. P-availability in the soil is restricted as it occurs mainly in insoluble forms. Global demand for Pi fertilizers is projected to increase significantly with the explosive growth of the global population. Thus, it has been predicted that global P-reserves will be depleted within 100 years or even sooner. Moreover, plants can use only a small amount of fertilizer P since 75–90% of added P is precipitated by metal–cation complexes and rapidly becomes fixed in soils. Plants have evolved with several mechanisms to cope with P-deficiency that include mycorrhizal symbiosis, phosphobacterial association, decreased growth rate, modification of root architecture for increased surface area, remobilization of internal inorganic phosphate, secretion of acid phosphatases (AP), exudation of organic acids, and enhanced expression of phosphate transporters. Such environmental concerns have led to the search for a sustainable way of P nutrition of crops. In this regard, phosphate-solubilizing microorganisms (PSM) have been seen as the best eco-friendly means for P nutrition of crop.

Keywords Mineral phosphate solubilization, P-fixation, Tri-calcium phosphate, Bacillus

1 Protocol

1. Suspend 1 g of rhizosphere soil scraped from the roots of each sample into sterile flask containing 100 ml of sterile deionized (DI) water to yield 10^2 dilutions.
2. Prepare a series of tenfold dilutions down to 10^{-9}. Every time vortex the tube for proper mixing of samples.
3. Plate 100 μL from each dilution onto Pikovskaya (PVK) media agar plates or National Botanical Research Institute's Phosphate growth medium (NBRIP) and incubate at 30 °C for up to 7 days [1]. The PVK medium contained (in g L^{-1}) 10 glucose, 0.5 yeast extract, 0.5$(NH_4)_2SO_4$, 0.1$MgSO_4 \cdot 7H_2O$, 5$Ca_3(PO_4)_2$, 0.2KCl, 0.002$MnSO_4 \cdot 2H_2O$, 0.002$FeSO_4 \cdot 7H_2O$, and 15 agar. NBRIP-agar medium contains glucose 10 g; tricalcium phosphate 5 g; magnesium chloride 5 g; magnesium sulfate ($MgSO_4 \cdot 7H_2O$) 0.25 g; potassium chloride

M. Senthilkumar, N. Amaresan and A. Sankaranarayanan, *Plant-Microbe Interactions: Laboratory Techniques*, Springer Protocols Handbooks, https://doi.org/10.1007/978-1-0716-1080-0_9,

0.20 g; $(NH_4)SO_4$ 0.10 g; agar 15 g; distilled water 1 L; the pH adjusted to 7.0 ± 0.2 before sterilization.

4. Wash the insoluble $Ca_3(PO_4)_2$ with DI water and centrifuge to remove soluble phosphate contaminants. Discard the supernatant, and dry the $Ca_3(PO_4)_2$ by using a vacuum-flask apparatus.
5. Select the colonies with clearing zones in the PVK agar plates and spot inoculate onto fresh PVK plates for better analysis of clearing zone formation.
6. Prepare and use modified version of the PVK to test the capability of the microorganisms to solubilize aluminum phosphate ($AlPO_4$) and iron phosphate ($FePO_4$). Modified medium contains all recipe of PVK except for the substitution of $Ca_3(PO_4)_2$ with either 5 g L^{-1} of $AlPO_4$ or 5 g L^{-1} of $FePO_4{\cdot}4H_2O$.
7. Quantitative determination of phosphate solubilizing activity on agar medium: The clearing zones formed by the bacteria on the respotted plates were quantified on the seventh day of incubation using the following equation:

$$\text{Phosphate solubilizing index} = \frac{\text{colony diameter} + \text{clearing zone}}{\text{colony diameter}} \times 100$$

2 Isolation of Halophilic P-Solubilizing Microbes from Marine

2.1 Protocol

1. Suspend 1 g marine sediment in 100 mL sterilized seawater and vortex it for 10 min.
2. Prepare tenfold dilutions of sediment suspension and spread 100 μL of dilutions onto a modified marine agar medium (1 L of seawater contains 5 g tryptone, 1 g yeast extract, and 15 g agar; pH 7.5), and incubate at 28 °C for 7 days.
3. Purify the bacterial colonies by re-streaking on 2216E agar plate and confirm the purity through microscopic observation.
4. Inoculate the pure bacterial isolates on the agar media supplemented either lecithin or $Ca_3(PO_4)_2$ and incubate at 28 °C for 1 week. The modified $Ca_3(PO_4)_2$ culture medium contained with the following ingredients: glucose 10 g, $(NH_4)_2SO_4$ 0.5 g, NaCl 30 g, KCl 0.3 g, $FeSO_4{\cdot}7H_2O$ 0.03 g, $MnSO_4{\cdot}4H_2O$ 0.03 g, $MgSO_4{\cdot}7H_2O$ 0.3 g, $Ca_3(PO_4)_2$ 10 g, agar 20 g, H_2O 1000 mL, pH 7.0–7.5. The lecithin culture medium was composed of: glucose 10 g, $(NH_4)_2SO_4$ 0.5 g, NaCl 30 g, KCl 0.3 g, $FeSO_4{\cdot}7H_2O$ 0.03 g, $MnSO_4{\cdot}4H_2O$ 0.03 g, $MgSO_4{\cdot}7H_2O$ 0.3 g, lecithin 0.2 g, $CaCO_3$ 5 g, yeast extract 0.4 g, agar 20 g, H_2O 1000 mL, pH 7.0–7.5.

5. Select the bacterial colonies with a clear phosphate solubilizing zone for further characterization/freeze-store in corresponding medium with 30% (v/v) glycerol at −80 °C.

Reference

1. Pikovskaya RI (1948) Mobilization of phosphorus in soil connection with the vital activity of some microbial species. Microbiologiya 17:362–370

Chapter 10

Estimation of Phosphate Solubilizing Capacity of Microorganisms

Abstract

Soluble orthophosphates combine with molybdate in acidic environment to form double salts. Though the amount of phosphomolybdate present may be sufficient to form precipitate, upon reduction, blue-colored soluble complex is formed, and the intensity of the blue color is proportional to the amount of phosphate present in the sample.

Keywords Phosphobacteria, Phosphomolybdate, Solubilization

1 Principle

Soluble orthophosphates combine with molybdate in acidic environment to form double salts. Though the amount of phosphormolybdate present may be sufficient to form precipitate, upon reduction, blue-colored soluble complex is formed, and the intensity of the blue color is proportional to the amount of phosphate present in the sample [1].

2 Procedure

1. Chlorostannous acid: 2.5 g of $SnCl_2 \cdot 2H_2O$ dissolved in 10 ml of concentrated HCl. Make the volume to 100 ml with distilled water.
2. Chloromolybdic acid: 15 g of ammonium molybdate dissolved in 400 ml of warm distilled water. Add 342 ml of 12 N HCl and cool it. Make up the volume into 1 l with distilled water.

3 Procedure

1. Prepare Pikovskya broth with known amount of inert phosphorous source like rock phosphate.

M. Senthilkumar, N. Amaresan and A. Sankaranarayanan, *Plant-Microbe Interactions: Laboratory Techniques*, Springer Protocols Handbooks, https://doi.org/10.1007/978-1-0716-1080-0_10,

2. Inoculate with test microorganisms and incubate at 28 °C on shaker for 3–4 days.
3. Centrifuge the broth in case of bacteria or filter the broth in case of fungi.
4. Take aliquot (0.1–1.0 ml) from the supernatant/filtrate and add 10 ml of chloromolybdic acid solution. Shake well and dilute the content of the flask into 45 ml.
5. Add 0.25 ml of chlorostannous acid and immediately make up the volume into 50 ml.
6. Measure the intensity of blue colored solution at 600 nm.
7. Find the corresponding amount of soluble phosphorous using standard curve.

Reference

1. Fiske C, Subbarow Y (1925) A colorimetric determination of phosphorus. J Biol Chem 66:375–400

Check for updates

Chapter 11

Isolation, Enumeration of Spores, and Inoculum Production of Arbuscular Mycorrhizal (AM) Fungi

Abstract

The mycorrhizal symbiosis with plants had been evolved 400 million years ago. Nearly 90% of plant species including flowering plants, bryophytes, and ferns can develop interdependent interactions with VAM. It forms vesicles, arbuscules, and hyphae in roots, and also spores and hyphae in the rhizosphere. Arbuscular mycorrhizae are formed only by fungi in the division Glomeromycota. Formation of hyphal network with plant roots significantly enhances the access of roots to a large soil surface area, causing improvement in nutrient supply and further plant growth. Mycorrhizal fungi can help the host plants to grow vigorously under stressful environment. A series of complex communication between the plant and the fungus results in enhanced photosynthetic rate and other gas exchange-related traits.

Keywords AM fungi, Spores, Wet sieving, Acid fuchsin–lactic acid stain

Wet sieving and decanting method of Gerdemann and Nicolson [1] is generally followed for quantitative estimation of arbuscular mycorrhizal (AM) propagules.

1 Procedure

1. Mix 250 g of soil in 1000 ml of water and allow heavier particles to settle down for few minutes.
2. Pour the liquid through course-sieve (500–800 m) to remove large pieces of organic matter.
3. Mix the filtrate with some more quantity of water, shake well, and allow heavier particle to settle for a few seconds.
4. Pass this suspension through a sieve fine enough to retain the desired spores generally 38–350μm. (*Note*: if the population is to be assessed, the finest available sieve should be employed. A nest of various sizes of sieves may be used instead of **steps 2–4**, however, care must be taken to ensure that the finest sieve does not over flow.)

M. Senthilkumar, N. Amaresan and A. Sankaranarayanan, *Plant-Microbe Interactions: Laboratory Techniques*, Springer Protocols Handbooks, https://doi.org/10.1007/978-1-0716-1080-0_11, © Springer Science+Business Media, LLC, part of Springer Nature 2021

5. Wash the material retained on the sieve to ensure that all colloidal material passes through the sieve.
6. Transfer the small amounts of remaining debris to a petri plate and examine under dissecting microscope. Microscopic counting of spores can be done with standard procedure. Spores of mycorrhizal fungi can be removed by Pasteur pipette fitted with a rubber bulb and subsequently used to establish pot cultures.

2 Mass Production of AM Fungi

1. Monocots like Sudan grass and Bahia grass with rapidly developing fibrous root system are ideal host plants for inoculums preparation of AM fungi. Spore/soil inoculums should be placed 2–3 cm below the seed of above grasses in pot culture.
2. Three to four months after inoculation, soil from pot culture should be sampled and checked for the maturity of new spores.
3. The entire pot can be harvested by pruning plants till soil level, removing the compact soil mass from the pots and chopping the mycorrhizal roots.
4. Before storage, the soil mixture should be air dried to the point at which there is no free water. After drying the culture should be packed in plastic bags sealed to prevent further drying and stored at 5 °C.

3 Quantification of AM Fungi Infection in Plant Roots

1. Collect sufficient terminal feeder roots attached to lower order roots from the plants.
2. Wash the root specimens with tap water and store it in FAA or process immediately.
3. Cut the roots into small pieces (5–10 mm) and boil it in a beaker containing 10% potassium hydroxide solution for 20 min to 1 h till it become soft in a well-ventilated exhaust hood. KOH solution clears the host cytoplasm and nuclei and further allows stain penetration.
4. Pour off the KOH solution and rinse three times with tap water or until no brown color appears in the rinsed water. Don't agitate the roots vigorously.
5. Cover the root in a beaker with alkaline H_2O_2 [(NaOH-3 ml; H_2O_2 (10%)-30 ml, water 567 ml] and keep it at room temperature for 10–20 min or until roots are bleached.

6. Rinse the roots in the beaker thoroughly using at least three complete changes of tap water to remove H_2O_2.
7. Cover the roots in the beaker with 1% HCl and soak for 3–4 min and then pour off the solution. Don't rinse the root samples after this step, because the specimens should be acidified for proper staining.
8. Cover the roots in the beaker with 0.01% trypan blue/acid fuchsin–lactic acid staining solution and warm it again for 10 min (Lactic acid solution: Lactic acid -875 ml, Glycerol-63 ml, Tap water-63 ml, Acid Fuchsin/trypan blue-0.1 g).
9. Remove excess of stain and cover the roots with above solution without acid fuchsin/trypan blue for distaining. Roots can be observed under microscopy.

Reference

1. Gerdemann JW, Nicolson TH (1963) Spores of mycorrhizal endogone species extracted from soil by wet sieving and decanting. Trans Br Mycol Soc 46:235–244

Chapter 12

Isolation of Zn Solubilizing Bacteria

Abstract

Zinc (Zn) deficiency is a well-known problem occurring in both plants and humans. Its deficiency in plants retards photosynthesis and nitrogen metabolism, causes the reduction in flowering and fruit development, decreases the synthesis of carbohydrates and phytohormones, and delays crop maturity leading to decrease in crop yield and nutritional quality of grains. Zinc solubilizing rhizobacteria includes the genera *Acinetobacter*, *Bacillus*, *Cyanobacteria*, *Gluconacetobacter*, *Pseudomonas*, and *Serratia*. This chapter deals with the techniques for isolation of zinc solubilizing bacteria [2].

Keywords Plant Growth Promoting Bacteria, Zn solubilization, Isolation, Clear zone

1 Materials

1. For the isolation of Zn-solubilizing isolates from rhizosphere soil, supplement NA with ZnO@0.629 g/l and D-glucose@10 g/l. Add cycloheximide@50 mg/l to all solid media to prevent fungal growth.

2 Methods

1. Uproot the plants from the field/pots, place them in the ice bag and transport to the laboratory and process within 5 h after sampling.
2. Remove the bulk soil from the plants by shaking and tapping the root systems gently.
3. Consider the soil that strongly adhere to the roots (i.e., a thin [<2-mm] soil layer) as rhizosphere soil.
4. Cut off the roots, including rhizosphere soil from the shoots, suspend in 30 ml of a sterile NaCl solution (9 g/l), and shake gently for 30 min (120 rpm). Prepare tenfold serial dilutions from the rhizosphere soil suspensions.

M. Senthilkumar, N. Amaresan and A. Sankaranarayanan, *Plant-Microbe Interactions: Laboratory Techniques*, Springer Protocols Handbooks, https://doi.org/10.1007/978-1-0716-1080-0_12,

5. Spread 100-μl aliquots of dilutions (10^{-3}, 10^{-4}, and 10^{-5}) on NA + ZnO.
6. Incubate the plates upside down at 28 °C for up to 14 days until halo-forming colonies appeared on the surfaces.

3 Observation

1. Select the colonies surrounded by clear clarification zones as ZSB, subculture the single colonies in NA plates for purification. For long-term storage, preserve the ZSB isolates at −80 °C in NB supplemented with 20% (vol/vol) glycerol.

4 Precautions/Note

1. For the isolation of P-Zn solubilizing bacteria, use modified Pikovskaya media [1] supplemented with Zn-P-tetrahydrate ($Zn_3(PO_4)_2{\cdot}4H_2O$). Incubation of plates was done at 30 ± 2 °C for 2 days. The halo zone colonies were selected and were purified by repeating subculture of particular single colonies on NB/LB plates.

References

1. Pikovskaya RI (1948) Mobilization of phosphorus in soil connection with the vital activity of some microbial species. Microbiologiya 17:362–370
2. Di Simine CD, Sayer JA, Gadd GM (1998) Solubilization of zinc phosphate by a strain of *Pseudomonas fluorescens* isolated from forest soil. Biol Fertile Soils 28:87–94

Chapter 13

Quantitative Estimation of Zn Solubilization Efficiency of Microorganisms

Abstract

Application of Zn fertilizers has been addressed to various crops but it transformed to various unavailable forms depending upon soil types and chemical reactions. The unavailable Zn compound can be converted back to available form through bio-augmentation of microbial inoculants having the ability to solubilize insoluble Zn compounds. This protocol deals with quantification of Zn solubilization.

Keywords Zn solubilization, Efficiency, Plant growth promoting bacteria

1 In Vitro Assessment of Zn Solubilization Efficiency

1.1 Method

1. We assessed solubilization efficiency by subjecting the strains to NA + ZnO and NA + $ZnCO_3$ media.
2. Spread 50 μl of the bacterial suspensions directly from fresh culture or glycerol stocks on NA plates.
3. After an overnight incubation at 28 °C, observe the NA plates for the development of thin layer of grown bacteria.
4. Transfer the bacterial lawn discs of 8 mm diameter upside down onto the surface of NA + ZnO and NA + $ZnCO_3$ plates with appropriate number of replicates per strain.
5. Incubate the inoculated NA + ZnO (ZnO@0.629 g/L) and NA + $ZnCO_3$ ($ZnCO_3$; 0.829 g/L) plates in the dark for 7 days at 28 °C. Calculate the Zn solubilization efficiency (SE) as follows [1].
6. Zn solubilization efficiency (SE) is defined as the average ratio from both ZnO and $ZnCO_3$ solubilization assays:

$$SE = ([AS/AG]_{ZnO} + [AS/AG]_{ZnCO3})/2,$$

where AS is the area of solubilization and AG is the area of bacterial growth. These areas were measured with ImageJ software (version 1.49).

M. Senthilkumar, N. Amaresan and A. Sankaranarayanan, *Plant-Microbe Interactions: Laboratory Techniques*, Springer Protocols Handbooks, https://doi.org/10.1007/978-1-0716-1080-0_13, © Springer Science+Business Media, LLC, part of Springer Nature 2021

2 In Vitro Liquid Assay for Determination of Zn Solubilization

2.1 Method

1. Dispense 50 ml of sterile solubilization assay medium (SAM) into 100-ml Erlenmeyer flasks for each bacterial strain, and subsequently inoculate with 2.5×10^8 CFU per ml. Conduct the experiment with appropriate number independent replicates per strain and incubate at 120 rpm 28 °C.
2. Prepare and incubate three uninoculated Erlenmeyer flasks of SAM without glucose and SAM with glucose under the same conditions and use as the controls.
3. Remove 8 ml aliquots from the Erlenmeyer flasks initially and after 1, 2, 3, 4, and 6 days of shaking at 120 rpm (28 °C), centrifuge at $2500 \times g$ for 10 min, and filter the supernatants through 0.22-μm-pore-size filters.
4. Measure the dissolved ZnO in cultures (i.e., solubilized Zn) and pH values at each time point with an inductively coupled plasma optical emission spectrometer (ICP-OES; Shimadzu ICPE-9820) and a pH meter (713 pH meter, electrode 6.0262.100; Metrohm), respectively.
5. Analyze the organic acids in sample filtrates by liquid chromatography (LC; NanoAcquity UPLC; Waters Corp.) coupled with a QTOF mass spectrometer (Synapt G2 HDMS; Waters Corp.). Use an ethylene-bridged hybrid (BEH) amide column (Acquity UPLC, column internal diameter, 200 μm; column length, 15 cm; particle size, 1.7 μm; Waters Corp.) for the separation of the organic acids.
6. Organic acids can also be measured in an uninoculated SAM without glucose control to determine the compounds originally present in the medium [2].

References

1. Nguyen C, Yan W, Letacon F, Lapeyrie F (1992) Genetic variability of phosphate solubilizing activity by monocaryotic and dicaryotic mycelia of the ectomycorrhizal fungus Laccaria bicolor (Maire) P.D. Orton. Plant Soil 143:193–199. https://doi.org/10.1007/BF00007873
2. Benjamin C, Laurie S, Emmanuel F, Cécile T (2018) Identification of heterotrophic zinc mobilization processes among bacterial strains isolated from wheat rhizosphere (*Triticum aestivum* L.). Appl Environ Microbiol 84(1): e01715–e01717

Chapter 14

Isolation of Sulfur-Oxidizing Bacteria

Abstract

Sulfur (S) is needed by plants and microorganisms for their growth. Speciation of S in soil is dependent on the chemical state of the soil, including redox potential and soil acidity. The S is oxidized in agricultural field soil by S-oxidizing bacteria (SoxB), mostly belong to the genus *Thiobacillus* that includes species like T. *thiooxidans*, *T. ferrooxidans*, *T. thioparus*, *T. denitrificans*, and *T. novellus*. The population density of these bacteria generally determines the degree to which S is converted to sulfate in soils. Population density of *Thiobacillus* can vary substantially in different soils. The population of S-oxidizing bacteria increases in the soil following application of S product. Most of the S in soils is bound to organic molecules, and therefore not readily available to plants. The requirement of S mainly differs between crop species and developmental stage of plants. Sulfur requirement is much greater for sunflower as compared to that for wheat and soybean. The S requirement is quite low during early growth stages for field bean, rice, and maize. Use of S oxidizers enhances the rate of natural oxidation of S and production of sulfates and makes them available to plants at their critical stages, resulting in increased plant yield.

Keywords SoxB, *Thiobacillus*, Sulfur, Oxidation

1 Materials

1. Starkey broth contains 3.0 g KH_2PO_4, 0.2 g $MgSO_4{\cdot}7H_2O$, 0.2 g $CaCl_2{\cdot}2H_2O$, 0.5 g $(NH_4)_2SO_4$, traces of $FeSO_4$ in 1000 ml distilled water with pH 8.0 [1].
2. NCL broth consists 0.2 g $(NH_4)_2SO_4$, 0.5 g $MgSO_4{\cdot}7H_2O$, 0.25 g $CaCl_2$ $2H_2O$, traces of $FeSO_4$, in 1000 ml of distilled water with pH 3–5.
3. Thiosulfate broth contains 5.0 g $Na_2S_2O_3$, 0.1 g K_2HPO_4, 0.2 g $NaHCO_3$, 0.1 g NH_4Cl in 1000 ml distilled water, with pH 8.0. Bromocresol purple was the indicator used.
4. Elemental sulfur at 10 g per liter was added to Starkey and NCL broths and half-an-hour steam sterilized for three consecutive days. Thiosulfate broth was devoid of S_0. For isolation of heterotrophic oxidizers add 5.0 g of glucose per liter of Starkey, NCL, and thiosulfate broths.

M. Senthilkumar, N. Amaresan and A. Sankaranarayanan, *Plant-Microbe Interactions: Laboratory Techniques*, Springer Protocols Handbooks, https://doi.org/10.1007/978-1-0716-1080-0_14,

2 Methods

1. Add 1 g of rhizosphere soil sample into 20 ml of the sterile broth dispensed in tubes, under aseptic conditions and incubate in BOD incubator at 32 °C for 25 days.
2. Take the inoculum from cultured broth and streak inoculate onto thiosulfate agar medium and incubate at 32 °C to obtain individual colonies.
3. Inoculate the individual colonies into Starkey broth with and without glucose having initial pH of 8.0 and incubated at 32 °C for 15 days. Based on the pH reduction of the broth with and without glucose, the isolates may be classified as chemoheterotrophs and chemoautotrophs.
4. Inoculate heterotrophic isolates in NCL and thiosulfate broth prepared with glucose. Inoculate autotrophic isolates in the same broths prepared without glucose and incubate at 32 °C for 15 days.
5. Purify the individual colonies by frequent transfer onto fresh broth at fortnightly intervals. Pick up and preserve the purified single colonies on thiosulfate slants.

Reference

1. Starkey RL, Collins VG (1923) Autotrophs. In: Norris JR, Ribbons DW (eds) Methods in microbiology, vol 38. Academic Press, New York, pp 55–73

Check for updates

Chapter 15

Enrichment Isolation of Sulfur-Oxidizing Bacteria

Abstract

Sulfur (S) is an essential plant nutrient; however, its content in the soil is only about 10% of that of the total N. Reduced S inputs from atmospheric depositions during recent years resulted in a negative S balance in agricultural soils, since crop plants have become increasingly dependent on the soil to supply crop need for S. Insufficient S availability leads to decreased yields and reduced S content in the plant products under extreme deficiency. Sulfur-oxidizing bacteria (SoxB) oxidize S into sulfate. The population of SoxB increases following application of S and this trait is used in enrichment isolation of SoxB.

Keywords Enrichment isolation, SoxB, Sulfur, Sulfate

1 Materials

1. The aerobic culture medium (ACM) for enrichment isolation of aerobic SOB contains the following components at the specified concentrations (in g/L): $^*Na_2S_2O_3$, 6.00; NaH_2PO_4, 1.22; Na_2HPO_4, 1.39; NH_4Cl, 1.00; $MgCl_2$, 0.10; $FeCl_3$, 0.03; $CaCl_2$, 0.03; $MnCl_2$, 0.03; KNO_3, 0.50; CH_3COONa, 1.00; $NaHCO_3$, 2.00.
2. The anaerobic culture medium (ANCM) for enrichment isolation of anaerobic phototrophic SOB contains (g/L): *$Na_2S_2O_3$, 10.00; NaH_2PO_4, 2.45; Na_2HPO_4, 2.78; NH_4Cl, 1.00; $MgCl_2$, 0.50; CH_3COONa, 1.00; sodium succinate, 1.00; $NaHCO_3$, 3.00; KNO_3, 0.50; NaCl, 1.00; yeast extract, 0.50; plus 1 mL of trace element solution which included (g/L): $FeSO_4 \cdot 4H_2O$, 1.80; $CoCl_2 \cdot 6H_2O$, 0.25; $CuCl_2 \cdot 2H_2O$, 0.01; $NiCl \cdot 6H_2O$, 0.01; $MnCl_2 \cdot 4H_2O$, 0.70; $ZnCl_2$, 0.10; H_3BO_3, 0.50; $(NH_4)_6Mo_7O_{24} \cdot 4H_2O$, 0.10; EDTA, 2.50; in 1 L distilled water.
3. *Because S_2will convert into H_2S and be lost as a gas from aerobic enrichment, use $Na_2S_2O_3$ instead of Na_2S as the sulfur substance. Add $Na_2S_2O_3$ aseptically from a sterile stock solution of 50 g/L $Na_2S_2O_3$ to the sterile media. Adjust the initial pH of media to 6.8. Add Agar (2%) as a solidifying agent [1].

M. Senthilkumar, N. Amaresan and A. Sankaranarayanan, *Plant-Microbe Interactions: Laboratory Techniques*, Springer Protocols Handbooks, https://doi.org/10.1007/978-1-0716-1080-0_15, © Springer Science+Business Media, LLC, part of Springer Nature 2021

2 Methods

1. Collect the activated sludge samples from a sulfide removing bioreactor at five different sites in sterile polyethylene bottles (500 mL volume), and mix together before being frozen in an ice cube box.
2. After transport to the laboratory, suspend 100 mL of the sludge samples in sterile phosphate buffer solution (PBS, pH 8.0) with the dilution ratio of 1:10 (W/V) and mix it by shaking with sterile glass beads.
3. After shaking for 30 min at 25 °C, allow mud particles, stones and sand to settle for 2 min and then collect the suspension.
4. Transfer the suspension into clean tubes, and centrifuge at 6000 × *g* for 5 min.
5. Resuspend the pellets twice in 100 mL of PBS (pH 8.0) and store at 4 °C until use for enrichments.
6. Enrichment protocol: Inoculate 10 mL of prepared sample into Erlenmeyer flasks containing 100 mL ACM medium and ampoule bottles containing 100 mL ANCM medium, respectively. Incubate the inoculated ACM for the enrichment for aerobic SOB at 30 °C and rotation speed of 180 rpm. For enrichment of SOB, frequently transfer 5% (V/V) culture fluid as the inoculums for each subculture. After four successive cultures, spread 0.1 mL of the enrichment and diluted samples (10^{-2}, 10^{-3}, 10^{-4}, and 10^{-5}) were spread onto agar ACM plates and incubate at 30 °C for 2–3 days. Pick up the colonies from ACM plates and inoculate in fresh fluid ACM to determine the sulfur-oxidizing ability.
7. For the enrichment of anaerobic and phototrophic SOB, incubate the inoculated ANCM liquid medium in an illumination incubator with light intensity of 2000 lux and temperature of 30 °C. Flush the ANCM culture medium with N_2 gas for 2 min to create an anaerobic condition. After enrichment, spread 0.1 mL of the culture sample and diluted samples (10^{-2}, 10^{-3}, 10^{-4}, and 10^{-5}) onto the agar ANCM plates and incubate at 30 °C and 2000 lux in an AnaeroPouchTM-Anaero system for 3–4 days. Pick up the colonies from plates and inoculate in fresh fluid ANCM to determine the sulfur-oxidizing ability.

Reference

1. Luo J, Tian G, Lin W (2013) Enrichment, isolation and identification of sulfur-oxidizing bacteria from sulfide removing bioreactor. J Environ Sci 25(7):1393–1399

Check for updates

Chapter 16

Quantitative Estimation of Sulfate Produced by Sulfur-Oxidizing Bacteria

Abstract

In soil, sulfur exists in oxidation states ranging from −2 to +6 in both organic and inorganic forms. The largest pool is organic S, which must undergo mineralization to inorganic S to become biologically available for microbial and plant uptake (mainly as sulfate). Alleviation of S deficiency in soils and increasing the S content of the organic pool is often achieved through addition of elemental S-fertilizer, which again must be oxidized by microorganisms before becoming plant available. Oxidation of reduced inorganic sulfur compounds including sulfide, sulfur, sulfite, and thiosulfate into sulfate is driven by an extensive range of photo- and chemolithotrophic sulfur-oxidizing prokaryotes, occupying very diverse, extreme and moderate environments and having evolved different pathways for sulfur oxidation. The pathway of the multienzymatic complex mediated by sox gene has been observed in a large group of prokaryotes (α-, β-, γ-, ε-Proteobacteria, Chlorobi, and Chloroflexi), being this system responsible for the oxidation of thiosulphate to sulfate. This chapter deals with the protocols on turbidimetric detection of sulfate as barium sulfate, atomic absorption spectrophotometer (AAS) technique with an air–acetylene flame and a barium (Ba) hollow cathode lamp, ion chromatography by IC with an electrical conductivity detector were discussed.

Keywords Sulfate, Quantification, Turbidimetric detection, Ion chromatography

1 Sulfate Detection by Spectrophotometry

1. Dilute the samples of culture supernatants@1:10 (v/v) with ultrapure water (MilliQ) and then add 1 mL of 6 M HCl solution containing 20 mg sulfur/L into 10 mL of sample.
2. After, add 500 mg of $BaCl_2{\cdot}2H_2O$ and allow the mixture to stand for 1 min.
3. Shake the tubes for 30 s to obtain complete dissolution of barium chloride crystals.
4. Finally, measure the turbidity in a spectrophotometer at 420 nm in 8 min maximum after the addition of crystals. To estimate the amount of sulfate in solution, prepare a standard

M. Senthilkumar, N. Amaresan and A. Sankaranarayanan, *Plant-Microbe Interactions: Laboratory Techniques*, Springer Protocols Handbooks, https://doi.org/10.1007/978-1-0716-1080-0_16,

curve at the concentration of 0, 2, 5, 10, 20, 25, 30 and 40 mg/L S-SO_4.

2 Sulfate Detection by Atomic Absorption Spectrometry (AAS)

1. Treat 1 mL of samples with 0.05 M HNO_3 and then dilute to 20 mL with ultrapure water (MilliQ).
2. Add 5 mL of 60 mg/L $BaCl_2$ solution into 5 mL of diluted samples homogenize and after 2 h, centrifuge at 3000 × *g* for 20 min.
3. Determine the remaining Ba in the solution by Fast Sequential Atomic Absorption Spectrometer with air-acetylene flame and barium (Ba) hollow cathode lamp.
4. The difference between the Ba concentration of the blank and the sample is proportional to the concentration of S-SO_4. For the determination of the amount of sulfate, develop a standard curve of S-SO_4, containing concentrations of 0, 1, 2, 4, 6, 8, 10, and 12 mg/L S-SO_4 in 0.05 M HNO_3.

3 Detection of Sulfate Ion Production

1. The amount of sulfate ion (SO_4^{2-}) produced during growth of sulfur-oxidizing bacteria on thiosulfate broth medium is measured spectrophotometrically.
2. Add barium chloride solution (10% w/v)@1:1 to bacterial culture supernatant and the mix vigorously. White turbidity due to barium sulfate formation indicates the positive result that can be measured with spectrophotometer at 450 nm. Compare the obtained data with the calibration curve using potassium sulfate (K_2SO_4) as standard for quantification [2].

4 Modified Method of Detection of Sulfate Ion Production

1. Add 1 mL of the sample/diluted sample to 1 mL of the conditioning reagent in a test tube and thoroughly mix it.
2. Add approximately 60 mg of crushed barium chloride crystals and vortex for 60 s at a constant speed. Determine the absorbance at 420 nm (A) of resulting suspension. A complete reaction mix excluding sulfate serve as blank.
3. A sample with a high content of suspended matter can be filtered or centrifuged prior to analysis since suspended matter in large amounts will interfere with the method.

4. The conditioning reagent contains 150 g NaCl, 100 mL glycerol (126 g), 60 mL concentrated HCl and 200 mL 95% ethanol and make up to 1 L with deionized water. Vigorous stirring is needed for the dissolution of the salt.
5. Sulfate calibration curve: Prepare standard sulfate solutions by dissolving K_2SO_4 in deionized water to known concentrations in the range 0 to 5 mM. Calibration curve using the standard solutions in this range can be fitted with a third-degree polynomial curve over the range 0 to 5 mM sulfate.

5 Sulfide Oxidase Assay

The enzyme activity is determined quantitatively by measuring the product of enzymatic reaction in the form of sulfate (SO_4^{2-}) [1] method.

1. The reaction mixture contains 4.5 mL of 0.1 M sodium acetate buffer (pH 5.6) and 1 mL supernatant, then 0.5 mL freshly prepared sodium sulfide (Na_2S) solution (0.06 g Na_2S, 0.16 g NaOH, 0.02 g EDTA $Na_2{\cdot}2H_2O$, 2 mL glycerol, and 40 mL distilled water).
2. Incubate the mixture at 30 °C for 30 min and stop the reaction by the addition of 1.5 mL NaOH (1.0 M) with mixing.
3. Assess the concentration of sulfate ion formed during sulfide oxidase assay through white turbidity by measuring absorbance at 450 nm using spectrophotometer. The amount of turbidity formed is proportional to the sulfate concentration in the sample. One unit of sulfide oxidase activity was defined as an amount of the enzyme required to produce 1 μmol sulfate/h/mL (U/mL).

References

1. Hirano T, Kurosawa H, Nakamura K, Amano Y (1996) Simultaneous removal of hydrogen sulphide and tri-methylamine by a bacterial deodorant. J Ferment Bioeng 81:337–342
2. Kolmert WMP, Hallberg KB (2000) A fast and simple turbidimetric method for the determination of sulfate in sulfate-reducing bacterial cultures. J Microbiol Methods 41:179–184

Part III

Techniques for Studying Abiotic Stress Management

Chapter 17

Enrichment Isolation of 1-Aminocyclopropane-1-Carboxylic Acid (ACC) Deaminase Producing Bacteria

Abstract

Plant growth promoting bacteria usually facilitate the development of plant growth by dropping the ethylene levels by deamination of 1-aminocyclopropane-1-carboxylic acid (ACC). The ACC is the direct precursor of ethylene production in plants. The enzyme called ACC deaminase, hydrolyses ACC into α-ketobutyrate and ammonia which can be further utilized by bacteria for their growth. Apart from decreasing the stress level of ethylene by ACC deaminase, the bacteria possessing this enzyme reported to increase the uptake of nutrients, plant resistance to various diseases, and enhanced production of various crops.

Keywords ACC deaminase, Ethylene stress, Salinity, Heavy metal, Pathogen attack

1 Principle

ACC deaminase producing bacteria reduce the level of ethylene stress and enable the plant to resist under biotic and abiotic stresses. Presence of 1-aminocyclopropane-1-carboxylate (ACC) deaminase activity in bacteria along with regulation of ACC, a precursor to plant ethylene levels, is one of the principal mechanisms for bacteria to provide beneficial effect on plants under abiotic stress. Therefore, the isolation of ACC deaminase producing bacteria act as potent plant growth promoters for plant and development under various stresses such as salt, flooding, ethylene, heavy metal, and pathogen attack [1].

2 Materials

1. Soil sample.
2. ACC (1-aminocyclopropane-1-carboxylic acid).

M. Senthilkumar, N. Amaresan and A. Sankaranarayanan, *Plant-Microbe Interactions: Laboratory Techniques*, Springer Protocols Handbooks, https://doi.org/10.1007/978-1-0716-1080-0_17,

3. DF (Dworkin and Foster) minimal salt media (salts per liter: 4.0 g KH_2PO_4, 6.0 g Na_2HPO_4, 0.2 g $MgSO_4 \cdot 7H_2O$, 2.0 g glucose, 2.0 g gluconic acid and 2.0 g citric acid with trace elements: 1 mg $FeSO_4 \cdot 7H_2O$, 10 mg H_3BO_3, 11.2 mg $MnSO_4 \cdot H_2O$, 124.6 mg $ZnSO_4 \cdot 7H_2O$, 78.22 mg $CuSO_4 \cdot 5H_2O$, 10 mg MoO_3, pH 7.2).
4. Rotary shaker.

3 Methodology

1. Weigh 1 g soil samples and inoculate in sterile DF minimal salt medium containing 3 mM ACC as the sole nitrogen source for the enrichment of ACC utilizing bacteria.
2. Incubate at 30 °C for 24 h on a rotary shaker at 200 rpm.
3. Make four fold dilutions of this culture and plate onto solid DF salt agar medium containing ACC (500 μmol mL^{-1}) and incubate it for 48 h at 30 °C.
4. After incubation, choose the bacterial colonies based on their colony morphology and purify it.
5. Maintain onto the respective medium slants at 4 °C and/or in 65% glycerol at −80 °C.

4 Note

1. Use freshly prepared solutions for the experiment.
2. Note down the observations carefully.

Reference

1. Bal HB, Nayak L, Das S, Adhya TK (2013) Isolation of ACC deaminase producing PGPR from rice rhizosphere and evaluating their plant growth promoting activity under salt stress. Plant Soil 366:93–105

Chapter 18

Estimation of ACC Deaminase Activity in Bacterial Cells

Abstract

Plant species require ethylene for seed germination and growth of seedling. Low amount of ethylene promote root initiation, growth, and root extension whereas, higher amount of ethylene inhibits root elongation. The bacterial enzyme 1-aminocyclopropane-1-carboxylic acid (ACC) promotes plant growth through the lowering ethylene levels to the plant. ACC exudes from plants root or seeds and is metabolized through microorganism which expresses ACC deaminase activity.

Keywords 1-Aminocyclopropane-1-carboxylic acid (ACC), Ethylene, PGP, α-Ketobutyrate

1 Principle

Plant species require ethylene for seed germination and growth of seedling. Low amount of ethylene promote root initiation, growth and root extension whereas, higher amount of ethylene inhibits root elongation. The synthesis of 1-aminocyclopropane-1-carboxylic acid (ACC) synthase enzyme catalyzes the conversion of *S*-adenosyl-L-methionine (SAM) to ACC, which is the immediate precursor of ethylene. High amount of ACC is produced by this reaction which is taken up by the bacterium and hydrolyzed to α-ketobutyrate and ammonia by ACC deaminase. The uptake and cleavage of ACC by ACC deaminase decreases the amount of ACC and ethylene, thereby acting as a sink for ACC [1–3].

2 Materials

- Tryptic soya broth (TSB)
- 0.1 M Tris-HCl
- DF minimal salt media (composition as mentioned in 17.2)
- DNF solution (0.1 g 2,4-dinitrophenylhydrazine in 100 ml of 2 N HCl)
- 1-aminocyclopropane-1-carboxylic acid (ACC)

M. Senthilkumar, N. Amaresan and A. Sankaranarayanan, *Plant-Microbe Interactions: Laboratory Techniques*, Springer Protocols Handbooks, https://doi.org/10.1007/978-1-0716-1080-0_18,

- α–ketobutyrate
- Eppendorf tubes
- Shaker
- Centrifuge

3 Methodology

1. Inoculate the bacterial culture in TSB and allow it to grow at 30 °C for 24 h on a shaker at 150 rpm.
2. Centrifuge the culture for 5 min at room temperature with maximum speed and discard the supernatant.
3. Wash the cell pellet with 0.1 M Tris-HCl (pH 7.6) and then suspend in DF minimal salt medium with 3 mM ACC as the sole nitrogen source.
4. Incubate the culture at 28 °C for another 36–72 h on a shaker. Harvest the bacterial cells by centrifugation at 5000 rpm for 5 min, wash twice with 0.1 M of Tris-HCl (pH 7.5), and resuspend in 200 μL of 0.1 M Tris-HCl (pH 8.5).
5. Add 5% toluene (v/v) and then vortex for 30 s to labilize the cells.
6. Incubate 50 μL of labilized cell suspension with 5 μL of 0.3 M ACC in an eppendorf tube at 28 °C for 30 min.
7. Prepare 50 μL of labilized cell suspension without ACC as a negative control, while 0.1 M Tris-HCl (pH 8.5) with 5 μL of 0.3 M ACC as a blank.
8. Mix the samples thoroughly with 500 μL of 0.56 N HCl by vortexing and remove the cell debris by centrifugation at 12,000 rpm for 5 min.
9. Transfer 500 μL supernatant to a glass test tube and mix 400 μL of 0.56 N HCl and 150 μL of DNF solution; and incubate the mixture at 28 °C for 30 min.
10. Add 1 mL of 2 N NaOH to the sample prior to measuring the absorbance at 540 nm.
11. Determine the concentration of α-ketobutyrate in each sample by comparing with a standard curve generated by α-ketobutyrate.

4 Notes

- Use freshly prepared solutions for the experiment.
- To determine the α-ketobutyrate concentration in the samples, prepare standard curve of α-ketobutyrate concentration (0.1–1.0 μmol).

References

1. Pandey S, Gupta S (2019) ACC deaminase producing bacteria with multifarious plant growth promoting traits alleviates salinity stress in French bean (*Phaseolus vulgaris*) plants. Front Microbiol 10:1506
2. Latif Khan A, Ahmed Halo B, Elyassi A, Ali S, Al-Hosni K, Hussain J, Lee IJ (2016) Indole acetic acid and ACC deaminase from endophytic bacteria improves the growth of *Solanum lycopersicum*. Electron J Biotechnol 19(3):58–64
3. Honma M, Shimomura T (1978) Metabolism of 1-aminocyclopropane-1-carboxylic acid. Agric Biol Chem 42(10):1825–1831

Check for updates

Chapter 19

PCR-Based Detection of ACC Deaminase Producing Bacteria

Abstract

The gaseous hormone ethylene synthesized in plant tissues from the precursor 1-aminocyclopropane-1-carboxylic acid (ACC) is involved in more than one physiological and developmental process in plants. Ethylene is important to regulate plants in terms of biotic and abiotic stress. The *acdS* gene encode for the ACC deaminase is used for the PCR-based detection of ACC deaminase producing bacteria.

Keywords PCR-based detection, *acdS* gene, 1-Aminocyclopropane-1-carboxylic acid (ACC), Ethylene

1 Principle

The *acdS* gene encoding ACC deaminase is used to study the phylogeny and diversity of ACC deaminase producers in bacteria and micro-eukaryotes (i.e., fungi and stramenopiles). In the root zone, the plant gets colonize by different types of *acdS*$^+$ microorganisms which contribute to degradation of ACC produced by roots, and the overall significance of ACC deamination for the plant is expected to result from the combined functioning of its *acdS*$^+$ microbial partners. Therefore, *acdS*-based PCR method is used to monitor *acdS* alleles.

2 Materials

- 0.8 and 1% agarose gel
- Forward and reverse primer
- 1× *Taq* buffer
- dNTPs
- *Taq* polymerase
- Thermocycler
- Electrophoretic apparatus
- UV-illuminator

M. Senthilkumar, N. Amaresan and A. Sankaranarayanan, *Plant-Microbe Interactions: Laboratory Techniques*, Springer Protocols Handbooks, https://doi.org/10.1007/978-1-0716-1080-0_19,

3 Methodology

1. Extract the genomic DNA from the ACC deaminase producing bacteria using phenol-chloroform extraction method or any other standard protocols.
2. Analyze the isolated genomic DNA on 0.8% agarose gel.
3. Use the isolated genomic DNA as a template to amplify the *acdS* gene by polymerase chain reaction (PCR) using specific primers.
4. Carry out the PCR reactions by preparing 25 μL reaction mixture containing 1× PCR reaction buffer, 2.5 mM dNTPs mixture, 10 pM of each primers, *Taq* DNA polymerase (1 U);

Name of primer	Primer sequence (forward and reverse)	References
AccF	5′-ATG AAT CTG AAT CGT TTT GAA C-3′ 5′-TCA GCC GTT GCG GAA CAG-3′	Ali et al. [1]
F primer ACC 4a R primer ACC 4b	5′-CAGCAGGAAAAGGATTTGGG-3′ 5′-ACTCCACTGAATTGAACCCG-3′	Wang et al. [2]
Primer F-*acdS* and R-*acdS*	5′-ATG AAY CTS CAR CGH TTY-3′ 5′-TYA RCC GTY SCG RAA RRT-3′	Onofre-Lemus et al. [3]
Primer NF-*acdS* and NR-*acdS*	5′-ATG AAY CTS CAR MRH TTY C-3′ 5′-TYA RCC GTY GCG RAA RAT V-3′	Onofre-Lemus et al. [3]

25 ng of template DNA.

5. Perform the PCR reaction in a PCR thermal cycler under the following conditions: initial denaturation at 94 ° C for 5 min, 35 cycles of denaturation at 94 ° C for 1 min, annealing at 54 ° C for 50 s and elongation at 72 ° C for 2 min; followed by final extension at 72 ° C for 7 min.
6. Separate the PCR product by electrophoresis through 1% agarose gel, purify and sequence it.
7. Compare the *acdS* sequence obtained with the existing database of *acdS*.
8. Construct the multiple-sequence alignments of *acdS* sequences with CLUSTALW2 software.
9. Infer the tree topology by the neighbor-joining method, and perform a distance matrix analysis using the program MEGA, version 7.1.

4 Notes

- Always use freshly prepared solutions to carry out PCR.
- Wear the hand gloves to avoid any hindrance in the protocol.
- Try to avoid the error occurring while pipetting the solutions.

References

1. Ali SZ, Sandhya V, Rao LV (2014) Isolation and characterization of drought-tolerant ACC deaminase and exopolysaccharide-producing fluorescent *Pseudomonas* sp. Ann Microbiol 64 (2):493–502
2. Wang C, Knill E, Glick BR, Défago G (2000) Effect of transferring 1-aminocyclopropane-1-carboxylic acid (ACC) deaminase genes into *Pseudomonas fluorescens* strain CHA0 and its gac A derivative CHA96 on their growth-promoting and disease-suppressive capacities. Can J Microbiol 46(10):898–907
3. Onofre-Lemus J, Hernández-Lucas I, Girard L, Caballero-Mellado J (2009) ACC (1-aminocyclopropane-1-carboxylate) deaminase activity, a widespread trait in *Burkholderia* species, and its growth-promoting effect on tomato plants. Appl Environ Microbiol 75(20):6581–6590

Check for updates

Chapter 20

Classical Triple Response Bioassay for Determining Stress Mitigating Potential of ACC Deaminase Producing Bacteria

Abstract

ACC deaminase producing bacteria can cleave the ethylene precursor ACC and thereby lower the level of ethylene in stressed plants. This reduces the inhibitory effect of high ethylene concentrations and allows plants to develop a better root system. Classical triple response bioassay is used to evaluate the presence of ethylene in a plant, to demonstrate whether the ACC deaminase producing bacteria is able to reduce the endogenous ethylene production or not.

Keywords ACC deaminase producing bacteria, Classical triple response bioassay, Ethylene, Stress mitigation, PGP

1 Principle

Environmental stresses are the most limiting factors that affect plant productivity and among these stresses, salinity is one of the most detrimental. Plants under salinity stress produce ethylene in high amount which can damage plants and cause epinasty, shorter roots, and premature senescence. An example of the negative effect of a high ethylene concentration is the classical triple response. This response consists of three distinct morphological changes in the shape of seedlings which includes inhibition of stem elongation, swelling of the stem, and a change in the direction of growth. Salt tolerance in plants depends mainly on the capability of root to decrease the ethylene level and the uptake of Na^+. Therefore, a check on the accelerated ethylene concentration and suppression of Na^+ uptake by plants could be helpful in minimizing salt stress. ACC-deaminase producing bacteria can cleave the ethylene precursor ACC and thereby lower the level of ethylene in stressed plants. This reduces the inhibitory effect of high ethylene concentrations and allows plants to develop a better root system [1]. Classical triple response bioassay is used to evaluate the presence of ethylene in a plant, to demonstrate whether the ACC deaminase producing

M. Senthilkumar, N. Amaresan and A. Sankaranarayanan, *Plant-Microbe Interactions: Laboratory Techniques*, Springer Protocols Handbooks, https://doi.org/10.1007/978-1-0716-1080-0_20,

bacteria is able to reduce the endogenous ethylene production or not [2].

2 Materials

- Seeds of a test plant
- ACC (1-aminocyclopropane-1-carboxylic acid)
- Sodium chloride to induce salt stress
- Green foil paper, airtight mason jars
- Cobalt (a chemical inhibitor of ethylene)
- ACC deaminase producing bacteria

3 Methodology

1. Grow the seeds of any plant in a 100 ml beaker and then expose it to 0, 3, 6, 9, and 12 mmol L^{-1} of ACC or 0, 3, 6, 9, and 12 dS m^{-1} salinity levels to provide stress condition.
2. Inoculate the seeds with ACC deaminase producing bacteria which would serve as a test.
3. Place the beakers in airtight mason jars and wrap it with green foil to provide green safe light.
4. Inoculate the seeds with ACC deaminase producing bacteria along with Co^{2+} (a chemical inhibitor of ethylene) in the presence of ACC (12 mmol L^{-1}) or salinity stress (12 dS m^{-1}) which would serve as a positive control.
5. Incubate for 7–10 days at room temperature.
6. Record the lengths and diameters of the seedlings after 7–10 days of incubation.

4 Notes

- Use freshly prepared solutions for the experiment.
- Analyze the results carefully.

References

1. Shaharoona B, Arshad M, Khalid A (2007) Differential response of etiolated pea seedlings to inoculation with rhizobacteria capable of utilizing 1-aminocyclopropane-1-carboxylate or L-methionine. J Microbiol 45(1):15–20
2. Nadeem SM, Zahir ZA, Naveed M, Asghar HN, Arshad M (2010) Rhizobacteria capable of producing ACC-deaminase may mitigate salt stress in wheat. Soil Sci Soc Am J 74(2):533–542

Chapter 21

Estimation of Ethylene in Plant-Bioassay System: Gas Chromatography

Abstract

Ethylene is seemed as a multifunctional phytohormone that regulates both growth and senescence. It promotes or inhibits growth and senescence strategies depending on its concentration, timing of application, and the plant species. Ethylene is the only olefin and exists inside the gaseous state below regular physiological conditions. Gas chromatography is a technique used for separation and analysis of volatile compounds with relatively high sensitivity.

Keywords Ethylene, Gas chromatography, Mitigation, Biotic–abiotic stresses

1 Principle

Plants regulate the biosynthesis of ethylene to control growth and development and to respond against biotic and abiotic stresses. To understand the molecular mechanism by which plants regulate ethylene biosynthesis as well as to identify stimuli triggering the alteration of ethylene production in plants, gas chromatography is used i.e., to measure ethylene concentration under *in vivo* condition. Gas chromatography is a technique used for separation and analysis of volatile compounds with relatively high sensitivity. Gas chromatography is used to measure the ethylene produced by plants, and become a valuable tool for ethylene research. Gas chromatography is a technique for analyzing compounds that are in vapor form or can be vaporized at an appropriate temperature [1, 2].

2 Materials

- Seeds of a test plant
- 30% bleach solution, 95% ethanol, and double distilled water
- Micro-centrifuge tubes
- Headspace vials and preassembled caps with septa

M. Senthilkumar, N. Amaresan and A. Sankaranarayanan, *Plant-Microbe Interactions: Laboratory Techniques*, Springer Protocols Handbooks, https://doi.org/10.1007/978-1-0716-1080-0_21,

- Aluminum foil
- Murashige and Skoog medium
- 0.6% top agarose
- A pipette filler, headspace crimper, and LAF hood
- A gas chromatograph with a headspace unit and column
- Carrier gases

3 Methodology

3.1 Surface Sterilization of Seeds

1. Place the seeds into a micro-centrifuge tube and add 900 μL of 95% (v/v) ethanol into the tube and incubate for 1 min at room temperature.
2. Discard the ethanol and add 30% (v/v) bleach solution and gently shake for 20 min at room temperature.
3. Discard the bleach solution and add 900 μL double distilled water (ddH_2O) to wash the seeds by gently inverting the tubes several times and repeat it for at least five times.
4. Discard the ddH_2O and add 1 mL of ddH_2O into the tube.

3.2 Preparation of Headspace GC Vials with Seeds

1. Sterilize headspace GC vials, preassembled caps with septa, and a pack of 2 × 2 in. precut aluminum foil using an autoclave.
2. Prepare MS media and 0.6% (w/v) top agarose and keep them in a 65 °C water bath and place the sterilized headspace GC vials on a rack that can securely hold the vials.
3. Prepare MS media with different cytokinin concentrations (0, 0.1, 0.5, and 1 μM), pour 3 mL of it into the headspace vials using pipette filler and allow the media to solidify.
4. Discard the ddH_2O from the tube of surface sterilized seeds. Then add 0.6% (w/v) top agarose to the tube and mix well using a pipette.
5. Withdraw 30–50 seeds mixed with 0.6% (w/v) agarose from the micro-centrifuge tube and place in the middle of the headspace vials and allow it to solidify for 10 min.
6. Seal the headspace vials with the sterilized aluminum foil and maintain the headspace vials at 4 °C for 2–4 days in the dark to stratify the seeds.
7. After stratification, bring vials to RT and remove the foil.
8. Using an electronic 20 mm automatic crimper (or manual crimper), crimp the headspace vials with the sterilized caps and place the vials in a plant growth chamber with dark conditions for 3 days.

3.3 Measurement of Ethylene with a Gas Chromatograph

1. Before turning the GC on, open gas valves to let gas flow into a GC.
2. Turn on the GC, headspace unit, software program for the headspace and GC operating system.
3. Prepare an ethylene standard curve using at least five different concentrations (0.1, 0.25, 0.5, 2.5, and 5 μL/L) of ethylene diluted from a known standard ethylene (10 μL/L).
4. Number the headspace vials, place them in the corresponding headspace unit and then run the GC.
5. Open the real-time running screen to identify ethylene peaks by finding the peak in the samples that has the same retention time as the standard ethylene peaks curve obtained.

3.4 Calculation of Ethylene Concentration from the Samples

1. Once GC run is finished, collect the sample from headspace vials, open them using a decapper and place them in a microwave until the MS agar in the vials gets melted.
2. Pour the seedlings from the vials into white weighing petri dish and record the number of seedlings per vial.
3. Retrieve the total concentration of the ethylene determined by comparing to the predetermined standard ethylene curve by opening the data files from the GC.
4. Divide the total concentration of ethylene with the number of seedlings and incubation days (or time), to get the unit of ethylene concentration (e.g., 10 μL/L per seedling per day).

4 Notes

- For obtaining the best analysis result, ultrapure gases are required.
- About 10–15 s of microwaving is enough for melting the MS agar in the vials because the long microwaving makes difficult to count the number of seedlings.

References

1. Yoon GM, Chen YC (2017) Gas chromatography-based ethylene measurement of Arabidopsis seedlings. In: Ethylene signaling. Humana Press, New York, pp 3–10
2. Yang SF, Hoffman NE (1984) Ethylene biosynthesis and its regulation in higher plants. Annu Rev Plant Physiol 35(1):155–189

Chapter 22

Estimation of Proline Content in Plant Tissues

Abstract

Accumulation of proline in higher plants is a sign of disturbed physiological condition, caused through stress condition. Free proline content can increase exposure of plants to biotic and abiotic stresses. Determination of free proline ranges is a useful assay to reveal physiological status and to evaluate the stress tolerance of higher plants. There are three methods available for the determination of free proline content. The isatin paper assay is easy and appropriate to assay proline content material in massive range of samples. The colorimetric dimension is quantitative and provides dependable information of the proline content material approximately. The HPLC-based totally amino acid analysis can be hired when attention of all amino acids has to be compared.

Keywords Proline, Reactive oxygen species (ROS), Isatin paper assay, Colorimetric assay, HPLC

The interaction of plant with both biotic and abiotic stress can trigger active defense reactions in plant. Reactive oxygen species (ROS) production in healthy plants is a common mechanism in which different by-products are produced by various metabolic pathways. Various environmental stresses lead to excessive production of ROS causing progressive oxidative damage and ultimately lead to cell death. Despite their destructive activity, they are well characterized second messengers in a variety of cellular process, including conferment of tolerance to various environmental stresses. To avoid ROS toxicity, plants produce enzymatic and nonenzymatic antioxidants that scavenge oxygen species. The first line in antioxidant defense is enzymatic antioxidant system which neutralizes superoxide anion (O_2^-) and hydrogen peroxide (H_2O_2). The enzymatic neutralization of ROS is ensured by antioxidant enzymes such as superoxide dismutase (SOD), catalase (CAT), guaiacol peroxidase (GPX), ascorbate peroxidase (APX), monodehydroascorbate reductase (MDHAR), dehydroascorbate reductase (DHAR), and glutathione reductase (GR), whereas, ascorbate (AsA), glutathione (GSH), carotenoids, tocopherols, and phenolic serves as a potent nonenzymatic antioxidants within the cell. The interactions between plant–microbe interactions rapidly generate the high amount of ROS and enhance plant tolerance to various

M. Senthilkumar, N. Amaresan and A. Sankaranarayanan, *Plant-Microbe Interactions: Laboratory Techniques*, Springer Protocols Handbooks, https://doi.org/10.1007/978-1-0716-1080-0_22,

abiotic stresses. The use of plant growth promoting bacteria (PGPB) as bioinoculants to mitigate stressful conditions represents a powerful technique in agricultural biotechnology for sustainable crop production.

The co-inoculation of plants with PGPB would minimize the deleterious effects caused by various environmental stresses by inducing enzymatic and nonenzymatic protection and maintaining the growth of the plant.

1 Estimation of Proline by Isatin Paper Assay

1.1 Principle

Isatin (1H-indole-2,3-dione) is a visualizing agents which gives different colors with different amino acids. It has been demonstrated that the reaction between isatin and proline forms a blue derivative called pyrrole blue. Based on the intensity of the blue color, the concentration of proline can be estimated [1, 2].

1.2 Materials

- Isatin solution (Isatin solution: Dissolve 10 g isatin in 1 L methanol and 5 mL of glacial acetic acid)
- Whatman filter paper (3 MM) or other chromatography papers (Impregnate the isatin solution with filter paper or chromatography paper and allow to air dry for 1 h. Papers should be stored in dark)
- 20% ethanol
- L-proline for preparation of standard proline solutions (0–5 mg/mL)
- Centrifuge
- Hot plate or Microwave oven

1.3 Methodology

1. Harvest the plant samples and measure their fresh weight.
2. Add 20% ethanol (10 μL/mg fresh weight) and grind the plant samples.
3. Centrifuge the samples for 5 min at room temperature with maximum speed and collect the supernatant.
4. Add 10 μL of supernatant to the isatin paper and dry the paper for 30 min.
5. Incubate the isatin paper with dried samples at 90 °C for 20 min to develop the blue color. Intensity of the blue color correlates with proline content.

1.4 Notes

- Use freshly prepared isatin solution for the experiment.

- For semiquantitative analysis use densitometry to generates the data. Freely available software such as ImageJ or NIH image softwares are suitable for densitometry and to generate the data.
- To determine the proline content in the samples, prepare different proline concentration (0–5 mg/mL) in 20% ethanol and perform the isatin reaction with 10 μL as mentioned above.

2 Estimation of Proline by Colorimetric Assay

2.1 Principle

The ninhydrin (2,2-dihydroxyindane-1,3-dione, CAS number 485-47-2) based colorimetric assay is extensively used to assay amino acids. At neutral pH ninhydrin destroys primary amino acids and also reacts with the released NH_3 to form a deep purple color known as Ruhemann's purple, which has a maximum absorption at 570 nm. Further in this pH, reaction with proline and other imino acids forms yellow orange product. In the case of low pH, the color is red with a peak of absorbance at 520 nm. At low pH also the purple is formed but rapidly loses an amine residue leads to colorless by-products. The levels of amino acids under stress condition are usually lower than proline levels [3, 4].

2.2 Materials

- 3% Sulfosalicylic acid (Dissolve 3 g of 5-sulfocalicylic acid (2-hydroxy-5-sulfobenzoic acid) in 80 mL distilled water and make up to 100 mL. At room temperature the solution can be stored for weeks.
- Acidic ninhydrin (Dissolve 1.25 g ninhydrin (1,2,3-indantrione monohydrate), 30 mL glacial acetic acid, and 20 mL of 6 M orthophosphoric acid by vortexing and with gently warming. Store the solution at 4 °C for 1 week.
- Toluene
- L-proline
- Centrifuge
- Cuvette
- Spectrophotometer

2.3 Methodology

1. Harvest the plant samples and weigh 100 mg for a reaction.
2. Add 3% sulfosalicylic acid (5 μL/mg fresh weight) and grind the sample. Perform the experiment on ice.
3. Centrifuge the sample at maximum speed for 5 min at room temperature.
4. Collect 100 μL of supernatant and prepare reaction mixture with 100 μL of 3% sulfosalicylic acid, 200 μL glacial acetic acid, 200 μL acidic ninhydrin. Mix the tube well.

5. Incubate the reaction mixture for 1 h at 96 °C.
6. After incubation terminate the reaction on ice.
7. Extract the samples with toluene by adding 1 mL of toluene to reaction mixture, vortex the sample for 30 s and leave on for 5 min to allow the separation of organic and water phases.
8. Remove the chromophore containing toluene into a fresh tube and measure the absorbance at 520 nm using toluene as reference.
9. Express the concentration as mg/g FW or micromole/g FW. For the determination of proline concentration standard concentration curve can be followed.

2.4 Note

- Use gloves to handle the tubes during extraction. Ninhydrin is very reactive.

References

1. Boctor FN (1971) An improved method for colorimetric determination of proline with isatin. Anal Biochem 43:66–70
2. Elliott RJ, Gardner DL (1976) Proline determination with isatin, in the presence of amino acids. Anal Biochem 70:268–273
3. Chinard FP (1952) Photometric estimation of proline and ornithine. J Biol Chem 199:91–95
4. Bates LS, Waldren RP, Teare ID (1973) Rapid determination of free proline for water-stress studies. Plant and Soil 39:205–207

Check for updates

Chapter 23

Determination of Glycine Betaine by Periodide Method

Abstract

Glycine betaine is an amphiphilic compound with two ends, hydrophobic positive end and a hydrophilic negative end. Glycine betaine is electrically neutral over a wide range of pH values. Glycine betaine compound is the most effective solute to improve drought and salinity tolerance in higher plants. Glycine betaine counteracts adverse consequences resulting from abiotic elements, the denaturation, and inactivation of proteins.

Keywords Glycine betaine, Periodide technique, Quaternary ammonium compound, Stress

1 Principle

Glycine betaine is a quaternary ammonium compound that accumulates in response to abiotic stress in various species. It can be determined by periodide technique spectrophotometrically. Glycine betaine reacts with iodine under acidic condition and at low temperature form betaine–periodide complex which appears in golden crystals that can be quantified by measuring optical density at 365 nm [1, 2].

2 Materials

- 2 N sulfuric acid
- Potassium triiodide (KI-I_2) (Add 15.7 g of iodine + 20 g of potassium iodide in 100 mL of sterilized distilled water).
- 1, 2–dichloroethane
- Glycine-betaine standard
- Liquid nitrogen
- Mortar and pestle
- Centrifuge

M. Senthilkumar, N. Amaresan and A. Sankaranarayanan, *Plant-Microbe Interactions: Laboratory Techniques*, Springer Protocols Handbooks, https://doi.org/10.1007/978-1-0716-1080-0_23,

- Cuvette
- Spectrophotometer

3 Methodology

1. Harvest fresh plant material and immediately lyophilize it in liquid nitrogen.
2. Collect lyophilized samples in prechilled mortar and grind it with pestle.
3. Add 1 mg of lyophilized cells in 1.5 mL of 2 N H_2SO_4.
4. Heat the mixture in water bath at 60 °C for 10 min to extract the quaternary ammonium compounds.
5. Centrifuge the mixture at 14,000 rpm for 10 min at room temperature.
6. Drawn off the supernatant in a new Eppendorf tube. Take 125 μL of supernatant sample and add 50 μL cold KI-I_2.
7. Incubate the tubes at 0–4 °C for 16 h.
8. Centrifuge the tubes at 14,000 rpm for 30 min at 0 °C.
9. Remove the supernatant carefully so that only crystals of glycine betaine remains attached to the wall and bottom of the tube. Maintain the temperature at 0 °C.
10. Dilute the precipitate into 1.4 mL of 1, 2–dichloroethane and incubate it for 2–2.5 h.
11. Read the absorbance in the spectrophotometer at 365 nm.
12. Prepare a standard of glycine betaine by preparing 1, 2, 4, 6, 8 μL of stock solution of betaine. Use these stock solutions instead of plant extract and perform as above.

4 Notes

- After the addition of potassium triiodide, perform remaining steps in darkened area because light degrades the color of the samples.
- Perform all the steps at mentioned temperature because glycine betaine crystals readily dissolved in the medium at temperature higher than 0 °C.

References

1. Valadez-Bustos MG, Aguado-Santacruz GA, Tiessen-Favier A, Robledo-Paz A, Munoz-Orozco A, Rascón-Cruz Q, Santacruz-Varela A (2016) A reliable method for spectrophotometric determination of glycine betaine in cell suspension and other systems. Anal Biochem 498:47–52
2. Grieve CM, Grattan SR (1983) Rapid assay for determination of water soluble quaternary ammonium compounds. Plant and Soil 70 (2):303–307

Chapter 24

Estimation of Betaine Aldehyde Dehydrogenase (BADH) Activity

Abstract

Betaine or glycine betaine is a quaternary ammonium compound accumulated by prokaryotes and eukaryotes in response to salinity or water stress condition. Betaine aldehyde dehydrogenase oxidizes the glycine betaine aldehyde in to the osmoprotectant glycine betaine under the osmotic stress. Many plant species, bacterial species, and marine animals accumulate betaine as osmoprotectant. It is biosynthesized from oxidation of betaine aldehyde reversibly by betaine aldehyde dehydrogenase.

Keywords Betaine aldehyde dehydrogenase, Osmoprotectant, Enzyme activity

1 Principle

Many plant species, bacterial species and marine animals accumulate betaine as osmoprotectant. It is biosynthesized from oxidation of betaine aldehyde reversibly by betaine aldehyde dehydrogenase.

$$\text{Betaine aldehyde} + H_2O + NAD^+ \rightarrow \text{Betaine} + 2H^+ + NADH$$

BADH is a highly specific and has higher affinity to reduce NAD^+ and $NADP^+$. The activity of BADH can be determined by considering the increase in optical density at 340 nm due to increase in NADH. One unit of enzyme activity is defined as the amount of the enzyme required for converting 1 μmole of NAD/-min/mg protein under the assay condition [1].

2 Materials

- Extraction buffer (50 mM HEPES-KOH (pH 8.0), 1.0 mM EDTA, 20 mM sodium metabisulfite, 10 mM sodium borate, 5.0 mM ascorbic acid and 5.0 mM dithiothreitol (DTT))
- Assay buffer (50 mM HEPES-KOH (pH 8.0), 1.0 mM EDTA, and 5.0 mM DTT)
- 1.0 mM NAD

M. Senthilkumar, N. Amaresan and A. Sankaranarayanan, *Plant-Microbe Interactions: Laboratory Techniques*, Springer Protocols Handbooks, https://doi.org/10.1007/978-1-0716-1080-0_24,

- 1.0 mM betaine aldehyde
- Sephadex G-25 column
- Mortar and pestle
- Centrifuge
- Cuvette
- Spectrophotometer

3 Methodology

1. Harvest 0.2–0.5 g of leaf material and grind it with 2 mL of extraction buffer with small amount of sand in mortar on ice.
2. Pellet out the debris by centrifuging at 10,000 rpm for 10 min at 4 °C.
3. Drawn off the supernatant and desalt it by using small columns of Sephadex G-25 which should be equilibrate with the assay buffer.
4. Perform the assay in 1.0 mL reaction mixture containing assay buffer, 1.0 mM NAD and the appropriate amount of enzyme extract.
5. Initiate the reaction by adding 1.0 mM betaine aldehyde.
6. Measure the increase in absorbance at 340 nm at 30 s interval over a period of 2 min.
7. Calculate the enzyme activity using extinction coefficient of NADH (6.22 mM^{-1} cm^{-1}).

4 Notes

- Perform the assay at 25 °C except for the extraction procedure (4 °C).
- Mixture of chemicals to be used as extraction buffer to combat inhibitory effect of phenols and polyphenol oxidase, which could be interfere the assay when simpler extraction buffer is used.
- Prepare the fresh extraction buffer.

Reference

1. Weretilnyk EA, Hanson AD (1988) Betaine aldehyde dehydrogenase polymorphism in spinach: genetic and biochemical characterization. Biochem Genet 26(1–2):143–151

Chapter 25

Estimation of Malondialdehyde (MDA) by Thiobarbituric Acid (TBA) Assay

Abstract

Lipid peroxidation is a complex processing in which polyunsaturated fatty acids (PUFAs) are subjected to attack via oxygen-derived free radicals resulting in the formation of lipid hydroperoxides. In biological tissues, these lipid hydroperoxides are broken-down from a variety of products such as aldehydes and ketones. MDA is a broadly used marker of oxidative lipid damage because of environmental stress.

Keywords Malondialdehyde, TBA assay, UV spectrophotometer

1 Principle

Malondialdehyde is the three carbon dialdehyde that are produced in plant species as the end product of polyunsaturated lipid peroxidation. This is most abundant aldehyde resulting from lipid peroxidation. For the determination of MDA, TBA assay is one of the widely used assays. In the TBA test reaction, one molecule of MDA reacts with two molecule of TBA which forms pink colored complex having absorption maximum at 532 nm in UV spectrophotometer [1, 2].

2 Materials

- Liquid nitrogen
- Trichloroacetic acid (TCA) (20% w/v) (Dissolve 20 g TCA in 100 mL of distilled water)
- Thiobarbituric acid (TBA) (0.67% w/v) (Dissolve 0.67 g TBA in 80 mL of distilled water) and heat at 50 °C for 45 min in a water bath and make the volume up to 100 mL
- MDA standard
- Water bath
- Ice bath

M. Senthilkumar, N. Amaresan and A. Sankaranarayanan, *Plant-Microbe Interactions: Laboratory Techniques*, Springer Protocols Handbooks, https://doi.org/10.1007/978-1-0716-1080-0_25,

- Centrifuge tubes and test tubes
- Cuvette
- Spectrophotometer

3 Methodology

1. Collect approximately 2 g of leaf sample and grind it in mortar and pestle with liquid nitrogen.
2. Add 0.1 g leaf tissue powder into centrifuge tube containing 1 mL of 0.1% TCA and mix by inverting the tube to precipitate the protein.
3. Centrifuge the homogenized samples at 12,000 rpm for 15 min.
4. Transfer the supernatant to a new test tube and mix with 4 mL of 20% TCA containing 0.67% TBA and mix well.
5. Boil the mixture at 95 °C for 15 min in water bath and quickly cool on ice bath for 10 min to stop the reaction.
6. Centrifuge the mixture at 10,000 rpm for 5 min and collect the supernatant.
7. Measure the optical density at 532 and 600 nm and calculate the concentration of MDA-TBA concentration based on the $\mathcal{E}$ value by comparing the standard curve.

   Where, $\mathcal{E}$ is the coefficient of absorbance ($1.53\ mM^{-1}\ cm^{-1}$)
8. To generate the standard curve, make a serial dilution of MDA stock solution to 1–10μM and treat the entire standard with TBA as described above.
9. Measure the optical density at 532 nm.
10. Generate the standard curve and calculate the MDA concentration according to following formula,

$$\text{MDA (nmol/mg protein)} = (A_{532} - A_{600}) \times V_r \times (V/V_t)/\mathcal{E} \times 1000/\text{Cp}$$

where,

A_{532}: Absorbance at 532 nm

A_{600}: Absorbance at 600 nm

V_r: Volume of reaction mixture

V: Total volume of crude enzyme solution

V_t: Volume of crude enzyme used in the test

Cp: Crude protein concentration (mg/mL)
$\mathcal{E}$: Extinction coefficient (1.53 $mM^{-1}cm^{-1}$)

4 Notes

- Perform all the steps at room temperature except the grinding of leaf tissue and extraction.
- To avoid the lipid oxidation during assay, butylated hydroxytoluene (0.01–2% BHT solution in ethanol) and EDTA (1 mM) can be added to the samples prior to addition of TCA.

References

1. Hagège D, Feutry S, Krsnik-Rasol M, Poder D, Menez JF (1995) Estimation of free and bound MDA in plant extracts: comparison between spectrophotometric and HPLC methods. In: Plant lipid metabolism. Springer, Dordrecht, pp 259–261
2. Chen T, Zhang B (2016) Measurements of proline and malondialdehyde content and antioxidant enzyme activities in leaves of drought stressed cotton. Bio-Protocol 6(17):1–14

Chapter 26

Estimation of Mono-Dehydroascorbate Reductase (MDAR) Activity

Abstract

Mono-dehydroascorbatereductase (MDAR) is an enzymatic compound of the ascorbate glutathione cycle in plants; this is one of the main antioxidant systems of plant cells for the protection against the damages caused by the reactive oxygen species (ROS). Ascorbate glutathione cycle is a metabolic pathway that detoxifies enzymatic, non-enzymatic antioxidants and hydrogen peroxidase.

Keywords Mono-dehydroascorbatereductase (MDAR), Ascorbate glutathione, ROS, UV spectrophotometer

1 Principle

Mono-dehydroascorbate reductase catalyzes reduction of mono-dehydroascorbate to ascorbate and involves the oxidation of NADH.

$$\text{Mono} - \text{dehydroascorbate} + 2\text{NADH} \rightarrow \text{Ascorbate} + 2\text{NAD}^{+}$$

The MDAR activity can be determined by measuring the oxidation of NADH using absorbance coefficient of 6.2 mM^{-1} cm^{-1} in UV spectrophotometer at 340 nm. One unit of enzyme activity is defined as the amount of enzyme required to oxidize 1.0 μmole of NADH/min/mg protein [1, 2].

2 Materials

- Potassium phosphate buffer (50 mM, pH 7.6)
- EDTA (1.0 mM)
- Sodium ascorbate (1.0 mM)
- Polyvinylpyrrolidone (4%)
- Ammonium sulfate (saturated in phosphate buffer)
- Tris-HCl (50 mM)

M. Senthilkumar, N. Amaresan and A. Sankaranarayanan, *Plant-Microbe Interactions: Laboratory Techniques*, Springer Protocols Handbooks, https://doi.org/10.1007/978-1-0716-1080-0_26,

- NADH (0.1 mM)
- Ascorbate (2.5 mM)
- Ascorbate oxidase
- Mortar and pestle
- Centrifuge
- Cuvette
- Spectrophotometer

3 Methodology

1. Take a prechilled mortar and add 0.5 g of fresh plant tissue, 0.2 g sand and grind it with 3.5 mL of 50 mM potassium phosphate buffer containing 1.0 mM EDTA, 1.0 mM sodium ascorbate, and 4% polyvinylpyrrolidone (PVP).
2. After grinding, add 1.5 mL of saturated $(NH_4)_2SO_4$ dissolved in phosphate buffer to precipitate the phenolic compound.
3. Centrifuge the contents at 13,000 rpm for 5 min at 4 °C.
4. Discard the residual and take the supernatant to carry out the assay.
5. Prepare a reaction mixture containing 50 mM Tris-HCl, 0.1 mM NADH, 2.5 mM ascorbate and supernatant in a final volume of 1 mL.
6. Initiate the reaction by adding ascorbate oxidase (0.14 units).
7. The reaction is followed by measuring the decrease in absorbance at 340 nm.

4 Notes

- Perform enzyme extraction at 0–4 °C and reaction at 25 °C.
- Use freshly prepared solutions at the time of assay.

References

1. Krivosheeva A, Tao DL, Ottander C, Wingsle G, Dube SL, Öquist G (1996) Cold acclimation and photoinhibition of photosynthesis in Scots pine. Planta 200(3):296–305
2. Hossain MA, Nakano Y, Asada K (1984) Monodehydroascorbate reductase in spinach chloroplasts and its participation in regeneration of ascorbate for scavenging hydrogen peroxide. Plant Cell Physiol 25(3):385–395

Check for updates

Chapter 27

Determination of Glutathione Activity

Abstract

Glutathione (GSH) is a tripeptide of glutamic acid, cysteine, and glycine synthesized in the cytosol. It is the principal component of the cellular redox buffering system. It is maintained in reduced form via an NADPH dependent reaction catalyzed by glutathione reductase. GSH is converted to oxidized glutathione (GSSG) upon oxidation via the formation of disulfide bond between cysteine residues. GSH is involved in the detoxification of many xenobiotics and overexpression of GR has been discovered in certain drug-resistant cancer cells.

Keywords Glutathione, Glutathione reductase (GR), Spectrophotometric method

1 Principle

GSH and GSSG are assayed according to the spectrophotometric method given by Smith [1]. Glutathione reacts with 5,5′-dithiobis (2-nitrobenzoic acid) to generate the 5-mercapto-2-nitrobenzoic acid (colored compound) and GSSG. The GSSG gets recycled to GSH (reduced form) by glutathione reductase enzyme used in the assay. The 5-mercapto-2-nitrobenzoic acid that gives color is the basis of the measure of total glutathione concentration by taking an absorbance at 412 nm in spectrophotometer.

2 Materials

- 5% sulfosalicylic acid (Add 5.0 g of sulfosalicylic acid in 100 mL of distilled water)
- 0.5 M potassium phosphate buffer (pH 7.5)
- 0.1 M sodium phosphate buffer (pH 7.5 with 5 mM EDTA)
- 2-Vinylpyrilidine
- 2 mM NADPH
- 6 mM 5,5′-dithiobis (2-nitrobenzoic acid) (DTNB)

M. Senthilkumar, N. Amaresan and A. Sankaranarayanan, *Plant-Microbe Interactions: Laboratory Techniques*, Springer Protocols Handbooks, https://doi.org/10.1007/978-1-0716-1080-0_27,

- Glutathione reductase (GR) (Dilute with 0.1 M phosphate buffer to get 1 unit/mL)
- GSH standards
- Mortar and pestle
- Centrifuge
- Cuvette
- Spectrophotometer

3 Methodology

1. Collect 1.0 g of fresh leaves with small amount of sand and homogenize it in 5 mL of 5% sulfosalicylic acid with mortar and pestle.
2. Add another 5 mL sulfosalicylic acid to it.
3. Sediment the insoluble material by centrifuging at 5000 rpm for 10 min.
4. Collect 1 mL supernatant and neutralize it with 1.5 mL of 0.5 M potassium phosphate buffer (pH 7.5). Use it for the total glutathione assay.
5. Take another 1 mL aliquot of supernatant and neutralize it as above. Add 0.2 mL of 2-vinylpyridine and mix well until the emulsion formed and incubate it at 25 °C for 1 h. Use this tube for the assay of GSSG.
6. Extract both the tubes twice with 5 mL diethylether. (Note: the reason for doing this extraction with diethylether is to extract 2-vinylpyridine which can slightly be inhibiting the enzyme assay.)
7. The reaction mixture for the assay contains 0.5 mL 0.1 M sodium phosphate buffer containing 5 mM EDTA, 0.2 mL 6 mM 5,5′-dithiobis(2-nitrobenzoic acid), 0.1 mL 2 mM NADPH and 0.1 mL (1 unit) glutathione reductase type III (from yeast).
8. Initiate the reaction by adding 0.1 mL enzyme extract and note down the optical density at 412 nm at every 15 s for 5 min.
9. For preparing standard curve, use the serially diluted GSH standards from 50 to 400 ng.
10. Add 0.1 mL GSH standards to the above reaction mixture and follow the same procedure as above.
11. The reaction mixture without NADPH and enzyme extract serves as blank.
12. An enzyme activity can be calculated based on standard curve analysis.

4 Notes

- All reagents should be prepared fresh.
- Assay temperature should be maintained 25 °C.

Reference

1. Smith IK (1985) Stimulation of glutathione synthesis in photorespiring plants by catalase inhibitors. Plant Physiol 79(4):1044–1047

Check for updates

Chapter 28

Estimation of Catalase

Abstract

Catalase is an important enzyme that dissociates hydrogen peroxide in to molecular oxygen and water. It is an intracellular enzyme, produced by prokaryotic and eukaryotic organisms. Catalase enzyme has been discovered in all aerobic and facultative anaerobes but it is not present in obligate anaerobes. Catalase activity is usually proportional to the amount of dissociation of hydrogen peroxide.

Keywords Catalase enzyme, UV spectrophotometer, H_2O_2 decomposition

1 Principle

Catalase enzyme catalyzes the following reaction of decomposition of hydrogen peroxide to give water and oxygen.

$$2H_2O_2 \rightarrow 2H_2O + O_2$$

Catalase activity can be measured by either following the decomposition of H_2O_2 or by the liberation of O_2. UV spectrophotometer is the method of choice for biological material. The decomposition of H_2O_2 by catalase can be followed directly by the decrease in extinction per unit time at 240 nm. The difference in extinction per unit time is the measure of catalase activity. One unit (U) of catalase activity is defined as the amount of enzyme that caused an absorbance change of 0.001/min under assay conditions [1].

2 Materials

- Phosphate buffer (PB) (50 mM), pH 7.0: (a) Dissolve 6.81 g KH_2PO_4 in 1000 mL of distilled water (b) dissolve 8.90 g $Na_2HPO_4 \cdot 2H_2O$ in 1000 mL of distilled water. Mix (a) and (b) in proportion of 1:1.5 (v/v).
- H_2O_2 (30 mM): Dilute 0.34 mL of 30% H_2O_2 with PB to 100 mL. Prepare freshly for each experimental assay. It can be stored at room temperature during assay.

M. Senthilkumar, N. Amaresan and A. Sankaranarayanan, *Plant-Microbe Interactions: Laboratory Techniques*, Springer Protocols Handbooks, https://doi.org/10.1007/978-1-0716-1080-0_28,

- Catalase solution: Commercially available lyophilized catalase powder can be dissolved in cold PB. Dissolve 10 mg of powder in 1 mL of cold PB. Dilute 5 μL catalase solution to 1 mL cold PB to obtain a solution with ~100 U/mL immediately before use.
- Blender.
- Centrifuge.
- Spectrophotometer.
- Cuvette.

3 Methodology

1. Collect the fresh plant tissue and homogenize it in a blender with ice-cold phosphate buffer.
2. Centrifuge the homogenate at 11,180 × *g* for 10 min at 4 °C and collect the supernatant.
3. Set up the spectrophotometer at 240 nm and calibrate it with mixture of 2 mL of phosphate buffer and 1 mL of 30 mM hydrogen peroxide in a 3 mL quartz cuvette as a blank.
4. In experimental cuvette, add 1 mL of PB with 1 mL of diluted sample. To start the assay, add 1 mL of 30 mM hydrogen peroxide to the cuvette and quickly place it in spectrophotometer to note the absorbance. (NOTE: The initial absorbance should be approximately 0.500–0.520. If required, add the PB to decrease the absorbance or add H_2O_2 to increase the absorbance.)
5. Mix with glass rod flattened at the end and immediately follow the decrease in absorbance with recorder for 3 min at every 30 s.
6. Use catalase solution as a control. Take 1.9 mL of PB in the cuvette, add 1 mL of 30 mM H_2O_2 and 0.1 mL of catalase solution (~10 U).
7. For setting blank, use 2 mL of PB and 1 mL of H_2O_2.
8. Calculate catalase activity with the following formula:

$$\text{U/mg} = (A_0 - A_{180}) \times V_t/\mathcal{E}_{240} \times d \times V_s \times C_t \times 0.001$$

where,

$(A_0 - A_{180})$ is the difference between the initial and final absorbance.

V_t is total volume of reaction (3 mL).

$\mathcal{E}_{240}$ is the molar extinction coefficient for H_2O_2 at OD_{240} (34.9 mol^{-1} cm^{-1}).

d is optical path length of cuvette (1 cm).

V_s is volume of sample (1 mL).

C_t is the total protein concentration in the sample.

0.001 is absorbance change that caused by 1 U of enzyme per min at 240 nm OD.

4 Notes

- For catalase assay, each test should be run one at a time.
- Temperature does not affect the assay, so it can be performed between 0–37 °C. But the preferred temperature is 20 °C.
- To avoid the bubbling, use low concentration of hydrogen peroxide.

Reference

1. Aebi H (1984) Catalase in vitro. In: Methods in enzymology. Academic, San Diego, pp 121–126

Check for updates

Chapter 29

Estimation of Superoxide Dismutase (SOD)

Abstract

Superoxide dismutase is a group of enzymes that catalyze the dismutation of superoxide radicals (O_2^-) to molecular oxygen and hydrogen peroxide supplying cellular protection against reactive oxygen species.

Keywords Superoxide dismutase, Superoxide, Hydroxylamine, Blue formazan

1 Principle

Superoxide dismutase (SOD) plays a vital role in scavenging superoxide (O_2^-) radicals. The activity of the enzyme can be determined by the inhibition in photoreduction of nitroblue tetrazolium (NBT) dye by superoxide radicals, which are generated by the autoxidation of hydroxylamine hydrochloride. Hydroxylamine hydrochloride is an intermediate in reduction of nitrate to ammonia and in oxidation of ammonia to nitrite and is autoxidized by superoxide radicals. Accompanying the autoxidation of hydroxylamine at high pH, NBT is reduced in the presence of EDTA which induces an increase in absorbance at 560 nm due to the accumulation of blue formazon. With the addition of SOD, superoxide radicals get trapped and hence there is an inhibition of reduction of NBT to blue formazon [1]. The enzymatic unit of superoxide dismutase has been defined as the amounts of the enzyme required to inhibit the reduction of chromogen by 50%.

2 Materials

- Phosphate buffer (PB) (50 mM), pH 7.0: Prepare as mentioned in Subheading 28.2.
- Sodium carbonate buffer (50 mM), pH 10.2.
- Nitroblue tetrazolium (96μM).
- EDTA (0.1 mM).

M. Senthilkumar, N. Amaresan and A. Sankaranarayanan, *Plant-Microbe Interactions: Laboratory Techniques*, Springer Protocols Handbooks, https://doi.org/10.1007/978-1-0716-1080-0_29,

- Hydroxylamine hydrochloride ($NH_2OH \cdot HCl$) 20 mM, pH 6.0.
- Triton X-100 (0.6%).
- Mortar and pestle.
- Centrifuge.
- Cuvette.
- Spectrophotometer.

3 Methodology

1. Grind 1.0 g of fresh leaf tissue in 10 mL of chilled 50 mM phosphate buffer in a prechilled mortar and pestle.
2. Centrifuge the homogenate at 11,180 × *g* for 10 min at 4 °C.
3. Use supernatant as an enzyme source.
4. Prepare a reaction mixture containing 1.3 mL of sodium carbonate buffer, 500μL of NBT and 100μL of Triton X-100 in test cuvette.
5. Set a blank without enzyme and NBT to calibrate the spectrophotometer.
6. Set another control having NBT but with no enzyme as reference control.
7. Add 100μL of hydroxylamine hydrochloride to initiate the reaction in the test cuvette.
8. After 2 min of incubation at room temperature, add 70μL of crude enzyme extract.
9. Read the absorbance at 560 nm immediately at every 15 s for 1–2 min.
10. Calculate the percentage inhibition in the rate of NBT reduction with the increase in absorbance at 560 nm.

$$\begin{aligned}&\%\text{Inhibition of NBT reduction by SOD}\\&\quad = \text{Control OD} - \text{Treatment OD}/\text{Control} \times 100\end{aligned}$$

Reference

1. Kono Y (1978) Generation of superoxide radical during autoxidation of hydroxylamine and an assay for superoxide dismutase. Arch Biochem Biophys 186:189–195

Check for updates

Chapter 30

Estimation of Ascorbate Peroxidase (APX)

Abstract

Ascorbate peroxidase (APX) is present in high amounts in chloroplast, cytosol, vacuole, and apoplastic space in leaf cell. Peroxidase is an enzyme observed in a wide range of organisms, from plants to human beings to bacteria. Its function is to breakdown hydrogen peroxide (H_2O_2), which is one of the toxic products produced as a by-product of the usage of oxygen for respiration.

Keywords Ascorbate peroxidase (APX), Hydrogen peroxide, Mono-dehydroascorbic acid (MDHA)

1 Principle

Ascorbate peroxidase (APX) is present in high amount in chloroplast, cytosol, vacuole, and apoplastic space in leaf cell. It catalyzes the following reaction

$$\text{Ascorbic acid} + H_2O_2 \rightarrow \text{Mono-dehydroascorbic acid} + H_2O$$

Ascorbate peroxidase reduces the hydrogen peroxide to water using ascorbic acid as a substrate and forms the oxidized form of mono-dehydroascorbic acid (MDHA). MDHA is metabolized to DHA spontaneously [1]. MDHA and DHA are converted to ascorbate by MDHA reductase and DHA reductase, respectively. The rate of oxidation of ascorbate can be followed by the decrease in absorbance at 290 nm spectrophotometrically. The enzyme activity can be calculated as μmol of ascorbic acid decomposed per minute with absorbance coefficient 2.8 mM^{-1} cm^{-1} at 290 nm [2].

2 Materials

- Potassium phosphate buffer (PPB) (50 mM, pH 7.0).
- Ascorbic acid (0.5 mM).
- Hydrogen peroxide (1.0 mM).
- EDTA (0.1 mM).

M. Senthilkumar, N. Amaresan and A. Sankaranarayanan, *Plant-Microbe Interactions: Laboratory Techniques*, Springer Protocols Handbooks, https://doi.org/10.1007/978-1-0716-1080-0_30,

- Mortar and pestle.
- Centrifuge.
- Cuvette.
- Spectrophotometer.

3 Methodology

1. Grind the fresh leaf tissue without midline in prechilled mortar and pestle with 50 mM potassium phosphate buffer.
2. Centrifuge the sample at 11,180 × *g* for 15 min at 4 °C.
3. Use the supernatant as a crude enzyme extract.
4. Prepare 0.9 mL of reaction mixture containing 50 mM PPB, 0.5 mM ascorbic acid, 1.0 mM H_2O_2, and 0.1 mM EDTA.
5. Add 0.1 mL of crude enzyme extract to initiate the reaction.
6. Wait for 30 s and follow the decrease in absorbance at 290 nm in UV spectrophotometer at interval of 15 s up to 60 s.
7. Use the reaction mixture without H_2O_2 and crude enzyme as reference.
8. Two methods can be used for calculating the enzyme activity.
   (a) Standard curve analysis
   - Find the initial and final concentration of ascorbic acid by using standard curve analysis. For preparing standard curve, use known concentrations of ascorbic acid in a reaction mixture instead of crude enzyme. Calculate the concentration of ascorbic acid oxidized/min/mg protein by following formula:

   $$\text{Quantity of ascorbic acid oxidized} = \text{Initial absorbance} - \text{Final absorbance}$$

   (b) Based on extinction coefficient
   - One unit of enzyme activity can be calculated as the amount of enzyme required to oxidize 1.0μmol of ascorbate/min/mg protein.

   $$\text{Unit activity (U/ min /mg)} = \Delta A_{290} \times V_r/\varepsilon \times V_t$$

   ΔA_{290}: change in the absorbance/min.
   V_r: total volume of reaction mixture in mL (1 mL).
   V_t: volume of sample taken in mL (0.1 mL).
   ε: extinction coefficient of ascorbate 2.8 mM^{-1} cm^{-1}.

4 Notes

- Hydrogen peroxide and ascorbate should be prepared in 50 mM potassium phosphate buffer to maintain the reaction mixture molarity.
- Buffer should be prepared fresh and store at room temperature.
- One may take fresh or old leaf sample but the consistency in taking same aged leaf should be maintained for each sample type.

References

1. Habib D, Chaudhary MF, Zia M (2014) The study of ascorbate peroxidase, catalase and peroxidase during in vitro regeneration of *Argyrolobium roseum*. Appl Biochem Biotechnol 172 (2):1070–1084
2. Nakano Y, Asada K (1981) Hydrogen peroxide is scavenged by ascorbate-specific peroxidase in spinach chloroplasts. Plant Cell Physiol 22 (5):867–880

Check for updates

Chapter 31

Estimation of Peroxidase (POD)

Abstract

Peroxidase includes in its widest sense a group of specific enzymes such as NAD-peroxidase, NADP-peroxidase, and fatty acid peroxidase. POD catalyzes the dehydrogenation of a large number of organic compounds such as hydroquinones, aromatic amines, phenols, *o*-cresol, *o*-toluidine, pyrogallol, leucomalachite green, guaiacol, benzidine, di-*o*-anisidine, etc.

Keywords Peroxidase, Guaiacol, Dehydrogenation product (GDHP), Guaiacol

1 Principle

Peroxidase can be determined either by decrease in hydrogen peroxide or by the formation of oxidized compound. Usually, many different substrates has been used because it catalyzes the dehydrogenation of many compounds such as hydroquinones, aromatic amines, phenols, *o*-cresol, *o*-toluidine, pyrogallol, leucomalachite green, guaiacol, benzidine, di-*o*-anisidine, etc. An accurate value is obtained with the use of di-*o*-anisidine but this compound should be limited because of its carcinogenicity. So, the generally used substrate for POD assay is guaiacol.

$$\text{Guaiacol} + H_2O_2 \rightarrow \text{Oxidized guaiacol} + 2H_2O$$

One mole of H_2O_2 oxidizes 1 mol of guaiacol and the end product is guaiacol dehydrogenation product (GDHP). The rate of formation of GDHP in the guaiacol assay is the measure of POD activity and this can be assayed by UV spectrophotometer at 436 nm wavelength. The extinction coefficient for GDHP at 436 nm is 6.39 $cm^2/\mu M$ guaiacol oxidized/μM H_2O_2 consumed [1].

M. Senthilkumar, N. Amaresan and A. Sankaranarayanan, *Plant-Microbe Interactions: Laboratory Techniques*, Springer Protocols Handbooks, https://doi.org/10.1007/978-1-0716-1080-0_31,

2 Materials

- Phosphate buffer (100 mM, pH 7.0): (a) Dissolve 13.61 g KH_2PO_4 in 1000 mL distilled water (b) 22.82 g $K_2HPO_4{\cdot}3H_2O$ in 1000 mL distilled water. Make 100 mL of buffer by mixing 39 mL solution a and 61 mL solution b.
- Guaiacol (20 mM): Dissolve 249 mg guaiacol in 100 mL distilled water.
- H_2O_2 (12.3 mM): Dilute 0.14 mL 30% H_2O_2 to 100 mL distilled water.
- Cuvette.
- UV spectrophotometer.

3 Methodology

1. Harvest 1.0 g of fresh plant tissue in 3 mL of 100 mM phosphate buffer and grind it in a precooled motor and pestle.
2. Centrifuge the homogenate at 16,099 × *g* for 15 min at 4 °C.
3. Use the supernatant for assay and discard the pellet.
4. Take 3.0 mL phosphate buffer, 50 μL guaiacol solution, 30 μL H_2O_2 solution reaction mixture as reference solution.
5. Add 100 μL of extracted supernatant to reference solution in the test cuvette and mix well and read the absorbance at 436 nm. Wait until the extinction has increased to 0.050.
6. Start the stopwatch and note down the time required for further increase in extinction of 0.100.
7. Note down the increase in extinction up to 3 min for better results.
8. Calculate the enzyme activity according to the following formula

$$\text{Enzyme activity (U/L)} = V_r \times V_t \times 1000/\varepsilon \times 1 \times \Delta t \times 0.100$$

where,

V_r: Total volume of reaction mixture (3.18 mL).

V_t: Crude enzyme solution in test cuvette (100 μL).

ε: Extinction coefficient at 436 nm (6.39 cm^2/μmol).

l: light path (1 cm).

Δt: time interval required for increase in extinction to 0.100.

0.100: One unit of POD defined as the amount of enzyme that increases 0.100 of absorbance at 436 nm/min.

4 Notes

- Guaiacol solution can be stored at −20 °C for couple of months.
- The extinction of 12.3 mM H_2O_2 solution should be 0.485 (±0.020) at 240 nm and 1 cm light path in UV spectrophotometer.
- Plant tissue extracted supernatant should be stored on ice during the assay.
- Most accurate values are obtained when the time required for extinction is between 1 and 3 min.

Reference

1. Pütter J (1974) Peroxidases. In: Methods of enzymatic analysis. Academic, San Diego, pp 685–690

Check for updates

Chapter 32

Estimation of Choline Content in Plant Tissues

Abstract

Choline is a source of methyl groups which is needed for many steps in metabolism. Choline being a constituent of lecithin might be found in every living cell. Choline is involved in phosphate as combination, possibly in sulfate translocation may act as a methyl donor and as a constituent of lecithin. It is the major component of membranes.

Keywords Choline, Free choline (FC), Phosphatidylcholine (PAC), Cytidine, Diphosphate choline, Glycerylphosphorylcholine (GPC)

1 Principle

Choline is commonly present in plant tissues as a free choline (FC), phosphatidylcholine (PAC), and some time as glyceryl phosphorylcholine (GPC), phosphorylcholine (PC), acetylcholine, sinapine and cytidine diphosphate choline, etc. Choline is involved in phosphate as combination, possibly sulfate translocation may act as a methyl donor and as a constituent of lecithin. It is the major component of membranes [1].

2 Materials

- Methanol.
- 0.1 M diethyl ether (Et_2O).
- 0.1 M HCl.
- 0.2 M potassium phosphate buffer.
- KI-I_2 (potassium iodide with iodine in water) solution: 15.7 g I_2, 20 g KI in 100 mL water.
- 1,2-dichloroethane.

M. Senthilkumar, N. Amaresan and A. Sankaranarayanan, *Plant-Microbe Interactions: Laboratory Techniques*, Springer Protocols Handbooks, https://doi.org/10.1007/978-1-0716-1080-0_32,

- Tapered glass centrifuge tube.
- Mortar and pestle.
- Centrifuge.

3 Methodology

3.1 Preparation of Extract

1. Grind the plant samples (0.2–1.5 g) with sand (0.5 g) and mix with 20 mL methanol.
2. Centrifuge the mixture at 447 × *g* for 5 min.
3. After centrifugation re-extract the residue with methanol for three times.
4. Evaporate the extraction and dissolve in 20 mL Et_2O-HCl (0.1 M) (1:1).
5. Allow phase to separate, add 10 mL Et_2O (Ethoxyethane) in combined extracted aqueous phase.
6. Add 10 mL HCl (0.1 M) to combined Et_2O phase extract and repeat for four times.
7. Evaporate the aqueous phase to dryness and dissolve the residue in water (200–500 mg FW/mL).

3.2 Assay Method

1. Dilute the extract with potassium phosphate buffer (0.2 M, pH 7.0) 1:1, for the precipitation of choline periodide.
2. Remove any precipitates by centrifugation (447 × *g* for 5 min).
3. Dilute the extracts and transfer to 12 mL tapered glass centrifuge tube and keep at 0 °C for 10 min.
4. Add 0.2 mL KI-I_2 reagent with gentle shaking and keep the tube at 0 °C for 50 min and scratch occasionally with glass rod to ensure complete crystallization of the choline periodides.
5. After centrifugation remove the supernatant, dissolve precipitate in 10 mL of 1,2-dichloroethane and read at 365 nm.
6. Quantify the choline concentration from standard curve prepared using choline.

4 Note

- Use freshly prepared solution.

Reference

1. Pearce RB, Strange RN, Smith H (1976) Glycinebetaine and choline in wheat: distribution and relation to infection by *Fusarium graminearum*. Phytochemistry 15(6):953–954

Chapter 33

Detection of Biofilm Forming Activity of Microorganisms

Abstract

Biofilm and EPS formation has associated with the capacity of bacteria to colonize plants in symbiotic, neutral, and pathogenic associations. Production of EPS helps in creating favorable environment for pathogen survival and growth inside the infected plant, thereby act as a protective barrier against plant metabolic defenses. Apart from the negative effects, EPS seems to influence cellular aggregation and biofilm formation on plant root surface there by promotes the plant growth. Other role of the EPS matrix includes their capacity to aggregate soil particles, a function that is important for soil structure, health, and fertility. The present unit describes various protocol used to isolate and characterize EPS/biofilm formation by bacteria.

Keywords Biofilm formation, EPS, Buried slide assay, Sand assay, Microtiter plate assay

Biofilms are derived from microbial communities which are characterized by abundant cells attached to a living or nonliving surface and surrounded in matrix of extracellular polymeric substances. The first report of biofilms was prepared using buried slide culture method to acquire an attachment of microorganisms. It has been demonstrated and validated that carbohydrate and proteins are the main components of EPS, however biochemical characterization of the EPS compounds remain unclear because of their complex structures and unique linkages. Therefore knowledge of the structure and functional properties of EPS is crucial for understanding the role of biofilms. There are numerous analytical techniques have been used to characterize the components and spatial distribution of EPS in biofilms. Presently spectroscopic and microscopic techniques are used for the isolation and characterization of EPS from different systems.

Biofilm and EPS formation has associated with the capacity of bacteria to colonize plants in symbiotic, neutral, and pathogenic associations. Production of EPS helps in creating favorable environment for pathogen survival and growth inside the infected plant, thereby act as a protective barrier against plant metabolic defenses. Apart from the negative effects EPS seems to influence cellular aggregation and biofilm formation on plant root surface

M. Senthilkumar, N. Amaresan and A. Sankaranarayanan, *Plant-Microbe Interactions: Laboratory Techniques*, Springer Protocols Handbooks, https://doi.org/10.1007/978-1-0716-1080-0_33,

there by promotes the plant growth. Other role of the EPS matrix includes their capacity to aggregate soil particles, a function that is important for soil structure, health, and fertility. The present unit describes various protocol used to isolate and characterize EPS/-biofilm formation by bacteria.

1 Buried Slide Assay

1.1 Principle

This assay describes the collection of soil biofilms on glass slides through direct contact between soil and microscopic slides. This method can be used in field or in the laboratory. This method was described by Parkinson et al. [1]. Note that the microbial population observed on the glass slide may not be similar to the undisturbed soil. The organisms that colonize the slide can be presumed to be soil microbes; however their proportion, relationship, and distribution of the microbes may be altered drastically.

1.2 Materials

- Sterile and cleaned glass slide (1 × 3 in.).
- Gram's staining kit/any other simple or differential stains.
- Forceps.
- Microscope.
- Spatula or long knife.
- Soil pots.

1.3 Methodology

1. Clean the microscopic slides with alcohol and sterilize by autoclaving or heat sterilize in Bunsen burner flame.
2. Select a soil sampling site and fill it in clay pot, plastic pot, or a plastic cup for placing microscopic slide.
3. Make slit in the soil core sample with the help of sterile knife or spatula.
4. Using sterile forceps insert one or more sterile slides into the slit.
5. Close the soil firmly around the slide to ensure that the soil is in contact with the slide.
6. Water the soil with little quantity.
7. Cover the pot/cup with plastic wrap and label it. If the experiment is carried out in the field, mark the site unnoticeably to attract unwanted attention.
8. Leave the slides for 1–3 weeks.
9. Remove the slides from the soil carefully by breaking the soil away from the slide.
10. Wipe the slide one side and clean with a tissue paper.

11. Stain the slide using Gram's stain or any other stain as per standard procedure for visualizing the soil biofilms.

1.4 Notes

- It is advised that staining without heat fixation is preferred. Heat fixation may dehydrate the biofilm that may lead to structure alteration.
- One may use slides having circular area to enumerate the cells from collected soils.
- For staining extracellular polysaccharide ruthenium red and Alcian blue are recommended.

2 Sand Assay

2.1 Principle

Biofilm forming bacteria are embedded in extracellular polymeric substance (EPS). The sand assay is used to determine biofilm forming ability of bacteria. Biofilm forming bacteria are able to form monolayer of cell on the sand.

2.2 Materials

- 24-well plate.
- Micropipette and tips.
- LB broth.
- M36 media plus citrate.
- Eppendorf tubes.
- Quartz sand.
- Citrate (0.4%).

2.3 Methodology

1. Grow bacteria in LB broth for overnight and sub culture in M36 plus citrate (0.4%) at a 1:5 dilution.
2. Place the sand in the bottom of the well in a 24-well plate and cover with bacterial suspension.
3. Keep the plate on shaker at room temperature for 4 h.
4. Remove the sand sample from each well and transfer into the eppendorf tube, wash five times with 500 μL M36.
5. Visualize under epifluorescent microscope that whether the wild type bacteria formed a monolayer of cells on sand or not.

2.3.1 For Quantification of Bacteria Attached to Sand

1. Add 50 μL of M36 into the sand containing tube.
2. Remove bacteria from the sand by alternating series of vortex (10 s) and sonication (10 s).
3. Remove 10 μL of suspension and use for plating.

3 Microtiter Plate Assay

3.1 Principle

Biofilm formation occurs when bacteria switch planktonic state to surface attached state. Microtiter plate assay is easy to detect the early stages of biofilm formation. Microtiter plate assay also used to study fungal biofilm formation. This method was first published in 1996 by Genevaux et al. [2] for rapid screening of *Escherichia coli* surface attachment.

3.2 Materials

- Microtiter plate.
- 0.1% crystal violet.
- Tray (with clean water) for rinsing step.
- Micropipette and tips.
- Paper towel or lab matt.
- Clean water.
- 30% Acetic acid.

3.3 Methodology

3.3.1 Preparing Microbial Biofilm

1. Incubate bacterial culture in an appropriate media (commonly used media for biofilm assays are LB and Mueller–Hinton broth (MHB)) at 37 ° C for 48 h.
2. Dilute the experiment culture at 1:100 in to fresh medium.
3. Add 100 μL diluted culture in to microtiter plate, and blank well containing media only.
4. Cover the microtiter plate with lid and incubate at appropriate temperature for 8–10 h.
5. After desired incubation time, remove planktonic cells by inverting microtiter plate.
6. Rinse the microtiter plate by deep in clean water tray (tray 1). While plate is deep in distilled water, gently rub the surface of the plate with gloved fingers to release bubbles and make sure that water enters all wells.
7. Keep the plate down face on the paper towel and remove excess water as much as possible.

3.3.2 Biofilm Staining

1. Add 125 μL 0.1% crystal violet in to all wells for 10–15 min at room temperature (ensure this volume covered the biofilm).
2. After 10 min invert the microtiter plate and remove the stains.
3. Submerge the plate in to clean water (tray 2), rub the entire surface of the plate, and ensure the water enters all wells (rinse at least twice).
4. Submerge the microtiter plate in tray 3 and rinse twice as mentioned above.

5. Remove the excess water by inverting plate on paper towel.
6. Allow it to air dry for some time until all excess water is evaporated.
7. The formation of violet rings indicates biofilm.

3.4 Notes

- Use freshly prepared solution.
- Tray 1 and tray 2 should contain clean water.
- Use media according to your testing organisms.

References

1. Parkinson D, Gray TRG, Williams ST (1971) Isolation of microorganism. In: Methods for studying the ecology of microorganisms, IBP handbook, vol 19. Blackwell Scientific Publications, Oxford, pp 36–55
2. Genevaux P, Muller S, Bauda P (1996) A rapid screening procedure to identify mini-Tn10 insertion mutants of *Escherichia coli* K-12 with altered adhesion properties. FEMS Microbiol Lett 142(1):27–30

Check for updates

Chapter 34

Extraction of Extracellular Polymeric Substances (EPS) from Bacteria

Abstract

Extracellular polymeric substances (EPS) are the vital part of the biomass. EPS may be of bacterial origin along with the bacterial lysis product or they are molecules from the wastewater adsorbed on bacterial cells. There are both physical and chemical methods for the extraction of the EPS from the bacteria. EPS are composed of diverse organic compounds and the composition of EPS depends on the nature of aggregates.

Keywords EPS, Biofilm, Extraction, Physical extraction, Chemical extraction

1 Principle

Extracellular polysaccharide substances (EPS) are the structural and functional components of biofilm. EPS is mainly lipid, protein, polysaccharide, and extracellular DNA (eDNA) [1]. The EPS matrix acts as a physical strength and protective barrier. The physical and chemical characteristics of the bacterial cells can be affected by EPS composition influencing factors such as cellular recognition, aggregation, and adhesion in their natural environments [2].

2 Materials

- Centrifuge.
- Sonicator.
- Refrigerator (4 °C).
- Ice cold isopropanol.
- CER (cation exchange resin).
- Hot water bath.
- 0.45 μM acetate cellulose membrane.

M. Senthilkumar, N. Amaresan and A. Sankaranarayanan, *Plant-Microbe Interactions: Laboratory Techniques*, Springer Protocols Handbooks, https://doi.org/10.1007/978-1-0716-1080-0_34,

- Deionized (DI) water.
- 2% (w/v) EDTA solution.
- 1 M NaOH.
- 0.22% formaldehyde.
- 8.5% sodium chloride.
- Dialysis membrane (3500 Da).

3 Methodology

3.1 Physical Extraction Protocols

3.1.1 Centrifugation

1. Inoculate the bacterial cultures in 50 mL of nutrient broth in a 250 mL conical flask.
2. Incubate the cultures on rotary shaker at 37 °C for 72 h.
3. Centrifuge the bacterial culture for 20 min at 11,180 × *g*.
4. After centrifugation, collect the supernatant and add twice the volume of ice cold isopropanol in it, store overnight at 4 °C.
5. After incubation, centrifuge the obtained precipitated material at 11,180 × *g* 20 min.
6. Air-dry the pellet and store at 4 °C for further use.

3.1.2 Sonication

1. Take 50 mL bacterial culture and sonicate at 37 W for 1 min using a sonicator probe.
2. Repeat the above steps (**steps 3–6** of Subheading 3.1.1 of centrifugation).

3.1.3 CER Extraction

1. Transfer 50 mL bacterial culture into the centrifuge tube (10 mL aliquot).
2. Add 2 g of hydrated CER and wash twice with phosphate buffer.
3. Vortex the sample at high capacity for 20 min.
4. Repeat the above steps (**steps 3–6** of Subheading 3.1.1 of centrifugation).

3.1.4 Heating Method

1. Heat the bacterial culture at 80 °C for 10 min.
2. Centrifuge the mixture at 11,180 × *g* for 10 min.
3. Filter the supernatant with 0.45 μM acetate cellulose membranes.
4. Finally, remove the extracting reagent in the EPS solution by membrane dialysis (3500 Da) for 24 h at 4 °C.

3.2 Chemical Extraction Protocols

3.2.1 Formaldehyde and Heating

1. Inoculate the bacterial cultures in 50 mL of nutrient broth in a 250 mL conical flask.
2. Incubate the flask on rotary shaker at 37 °C for 72 h.
3. Centrifuge the bacterial culture for 20 min at 11,180 × *g* and harvest the pellet.
4. Resuspend the freshly harvested cell pellet in 10 mL 0.22% formaldehyde in 8.5% sodium chloride and incubate for 2 h at 48 °C.
5. Centrifuge the mixture of suspension at 11,180 × *g* for 20 min and resuspend the resulting pellet containing the EPS in 10 mL deionized (DI) water.
6. Centrifuge the suspension again at 11,180 × *g* for 10 min to rinse away any remaining cellular material, collect the pellet and measure the weight.

3.2.2 EDTA Method

1. Take 30 mL of 2% (w/v) EDTA solution in deionized water (DI) and add 30 mL bacterial culture.
2. Stir the mixture for 3 h at 4 °C.
3. Centrifuge the mixture at 11,180 × *g* for 20 min (4 °C).
4. Filter the supernatant with 0.45 μM acetate cellulose membranes.
5. Remove the extracting reagent in the EPS solution by membrane dialysis (3500 Da) for 24 h at 4 °C.

3.2.3 NaOH Extraction

1. Add 12 mL of 1 M NaOH in to 30 mL biofilm suspension.
2. The suspension is stirred for 3 h at 4 °C.
3. Centrifuge the suspension at 11,180 × *g* for 20 min and filter the supernatant with 0.45 μM acetate cellulose membranes.
4. Remove the extracting reagent in the EPS solution by membrane dialysis (3500 Da) for 24 h at 4 °C.

4 Note

- Use freshly prepared chemicals.

References

1. Fang HH, Jia XS (1996) Extraction of extracellular polymer from anaerobic sludges. Biotechnol Tech 10(11):803–808
2. Dignac MF, Urbain V, Rybacki D, Bruchet A, Snidaro D, Scribe P (1998) Chemical description of extracellular polymers: implication on activated sludge floc structure. Water Sci Technol 38:45–53

Check for updates

Chapter 35

Microscopic Technique for Studying Biofilm Formation

Abstract

Biofilms are groups of microbes attached to the surface, which can be observed in clinical, industrial, and natural settings. In fact, existence in a biofilm probably represents the predominant mode of growth for microbes in most environments. Mature biofilms have a few distinct characteristics. Biofilm microbes are commonly surrounded through an extracellular matrix that offers structure and protection to the community. Fluorescently labeled lectins have been used in mixture with epiflourescence microscopy and confocal laser scanning microscopy to permit the visualization and characterization of carbohydrate containing EPS in biofilm.

Keywords Biofilm, EPS, Confocal laser microscopy, Fluorescently labeled lectins

1 Principle

Single or multispecies communities together make microbial biofilm that accumulate interface (solid–liquid), where the microorganisms live at high densities in a matrix of hydrated EPS. The main role of EPS matrix is to give physical strength to biofilm such as attachment, mechanical strength, and antibiotic resistance [1].

2 Materials

- Two-chamber glass slide.
- Stain: (1) FITC-labeled Hippeastrum hybrid lectin (HHA) (2) Hoechst 33342.
- Fluorescent microscope.
- PBS solution (2×).

3 Methodology

1. Take two chamber slides with 1 mL of tryptic soya broth.

M. Senthilkumar, N. Amaresan and A. Sankaranarayanan, *Plant-Microbe Interactions: Laboratory Techniques*, Springer Protocols Handbooks, https://doi.org/10.1007/978-1-0716-1080-0_35,

2. Grow microbial culture in to two-well chamber slide for 1–3 days.
3. Wash the biofilm wells with PBS (2×) solution.
4. Stain the biofilm chamber with FITC-labeled HHA and Hoechst 33342 (a nucleic acid dye).
5. After staining, keep it in PBS for 1 h at room temperature.
6. After incubation wash it with PBS (2×) solution.
7. Remove the glass slide and observe attached biofilm using fluorescent microscope or scanning confocal microscope.
8. Observe the images using NIS-Elements or ZEN (Zeiss) software packages.

4 Notes

- The FITC labeled mannose specific HHA lectin—stain exopolysaccharides (green).
- Hoechst 33342—stain the bacterial nucleic acid (blue).
- 1-day-old microbial biofilm should be used.

Reference

1. Strathmann M, Wingender J, Flemming HC (2002) Application of fluorescently labelled lectins for the visualization and biochemical characterization of polysaccharides in biofilms of *Pseudomonas aeruginosa*. J Microbiol Methods 50(3):237–248

Check for updates

Chapter 36

Determination of Carotenes

Abstract

In 1883, Borodin discovered that the carotenoid pigments could be separated in to alcohol and petroleum ether soluble groups. The dilatory carotenes or beta-carotenes convert in to vitamin A (retinol). The beta-carotene is a precursor of vitamin A. Different types of methods are available to detect the carotenoids, such as thin layer chromatography, high pressure liquid chromatography, and MALDI-TOF. Most commonly HPLC in combination with UV-Vis absorption method is used for identification and quantification of carotenoids.

Keywords Carotenoids, Carotenes, Plant tissue, Chromatography, HPLC

1 Principle

Carotenoids are tetraterpenoid compounds which are commonly present in plants. Carotenoids occurring in different chemical forms have characteristic features and functions [1]. On the basis of solubility, carotenoids are extracted and partitioned in organic solvents. The separation of individual components is effected by chromatography on activated magnesia [2, 3].

2 Materials

- Spectrophotometer.
- Chromatographic column.
- Mortar and pestle.
- Volumetric flask.
- Petroleum ether.
- Chloroform.
- Sodium sulfate adsorbent: Mix one part of magnesium oxide (MgO) with three parts of supercel.

M. Senthilkumar, N. Amaresan and A. Sankaranarayanan, *Plant-Microbe Interactions: Laboratory Techniques*, Springer Protocols Handbooks, https://doi.org/10.1007/978-1-0716-1080-0_36,

- 3% acetone in petroleum ether.
- Sea sand, purified.
- Surgical cotton.

3 Methodology

1. Grind 2–5 g plant sample using mortar and pestle or grind it using food chopper with 40 mL acetone, 60 mL hexane, and 0.1 g $MgCO_3$ and blend it for 5 min.
2. Filter the suspension and collect the residue.
3. Wash the collected residue two times using 25 mL of acetone and 25 mL hexane and mix both the extracts together.
4. Wash acetone from extract with 500 mL portion H_2O.
5. Transfer upper layer to 100 mL of volumetric flask containing 9 mL acetone and dilute to volume with hexane.
6. Prepare chromatographic tube which is packed with magnesia-diatomaceous earth (1:1) concentration.
7. Prepare column using small glass wool or cotton plug inside the tube, add loose adsorbent to 15 cm depth.
8. Pour the extract into the column with vacuum continuously applied to flask.
9. Develop chromatogram using 50 mL acetone–hexane (1:9) and wash visible carotenes through adsorbent. Keep the top of the column covered with solvent during entire operation.
10. Collect entire eluate. Transfer the eluate to 100 mL volumetric flask and dilute the volume with acetone–hexane mixture.
11. Determine the absorbance at 436 nm with spectrophotometer. Use high purity β-carotene to calibrate the instrument.
12. Prepare calibration curve and calculate carotene content (mg/100 g).

4 Notes

- β-carotene is unstable in light and sensitive to air oxidation. The extraction should be processed by avoiding light and oxidation.
- Never handle petroleum ether near a flame.
- Avoid inhaling the solvents directly.

References

1. Russell WC, Taylor MW, Chichester DF (1935) Colorimetric determination of carotene in plant tissue. Plant Physiol 10:325–340
2. Müller H (1997) Determination of the carotenoid content in selected vegetables and fruit by HPLC and photodiode array detection. Z Lebensm Unters Forsch 204:88–94
3. Oliver J, Palou A (2000) Chromatographic determination of carotenoids in foods. J Chromatogr A 881:543

Chapter 37

Determination of Chlorophyll

Abstract

Chlorophyll is the major pigment in plants and acts as a biomass indicator of aquatic microalgae which assists food webs within the sea. The chlorophyll content is dependent on the environment and season. In phytoplankton, low chlorophyll is observed during the winter season, and this may be due to limitation of sunlight. There are some methods to estimate the content of chlorophyll, such as based on the absorption of light by aqueous acetone extracts, based on the absorbance or reflectance of certain wavelengths of light by intact leaves using handheld chlorophyll meters.

Keywords Chlorophyll, Pigments, Spectrophotometer, Aqueous acetone extracts

1 Principle

The chlorophylls are essential components for photosynthesis and occur in chloroplasts as green pigments in all photosynthetic plant tissues. Chlorophyll is extracted in 80% acetone and the absorption is read at 643, 645, and 663 nm in spectrophotometer. Absorption coefficient is used to calculate the amount of chlorophyll [1, 2].

2 Materials

- Acetone (80% acetone: 80 mL in 20 mL of distilled water).
- Centrifuge.
- Mortar and pestle.
- Volumetric flask.
- Spectrophotometer.

3 Methodology

1. Harvest 1 g of leaf sample and chopped it into small pieces.

M. Senthilkumar, N. Amaresan and A. Sankaranarayanan, *Plant-Microbe Interactions: Laboratory Techniques*, Springer Protocols Handbooks, https://doi.org/10.1007/978-1-0716-1080-0_37,

2. Grind the leaf samples in sterile mortar and pestle into fine pulp by addition of 20 mL of 80% acetone.
3. Centrifuge the mixture at 2795 × *g* for 5 min and transfer the supernatant to a 100 mL volumetric flask.
4. Repeat the procedure till residue of supernatant becomes colorless.
5. Make the final volume to 100 mL by addition of 80% acetone.
6. Read the absorbance at 645 and 663 nm against the solvent blank.
7. Calculate the amount of chlorophyll in per gram of sample using following equation

$$\text{Chla} = (12.7 \times \text{A663} - 2.69 \times \text{A645}) \times V/(1000 \times w)$$

$$\text{Chlb} = (22.9 \times \text{A643} - 4.68 \times \text{A665}) \times V/(1000 \times w)$$

$$\text{Total chlorophyll} = (20.2 \times \text{O.D.at 645 nm}) + (8.02 \times \text{O.D.at 663 nm})$$

where V is the total extract volume and w is the leaf fresh weight.

4 Note

- Dimethyl sulfoxide (DMSO) can also be used in place of acetone and the experiment can be performed as per above method.

References

1. Arnon DI (1949) Copper enzymes in isolated chloroplasts. Polyphenoloxidase in *Beta vulgaris*. Plant Physiol 24:1–15
2. Lorenzen CJ (1967) Determination of chlorophyll and pheo-pigments: spectrophotometric equations. Limnol Oceanogr 12:343–346

Chapter 38

Determination of Chlorophyll Fluorescence

Abstract

Chlorophyll florescence (CF) is an indirect indicator of the physiological status of chlorophyll and has the benefit of detecting cellular damage as consequence of senescence in advance of the development of visible signs and symptoms. Chlorophyll florescence analysis is one of the most powerful techniques available to the plant physiologists and ecophysiologists.

Keywords Chlorophyll florescence, Photochemical efficiency, Photochemical quenching, PSII reaction

1 Principle

Chlorophyll fluorescence is the most widely used method in photosynthesis and plant stress research. Photochemical efficiency of photosystem (PS)-II is measured as chlorophyll fluorescence in terms of F_v/F_m ratio (variable fluorescence/maximum fluorescence). The extent of photo inhibition induced by any environmental stress can be rapidly assessed by measuring the maximum photochemical efficiency of PSII [1].

2 Materials

1. Portable fluorometer.

3 Methodology

1. Collect leaves from the replicate plants.
2. Keep the leaves in the holder of the fluorometer and calculate minimum chlorophyll flurorescence (F_o).
3. Expose leaves to 0.6 s saturating pulse of light and calculate value of maximum chlorophyll flurorescence (F_m).
4. Calculate variable fluorescence (F_v) as $F_m - F_o$.

M. Senthilkumar, N. Amaresan and A. Sankaranarayanan, *Plant-Microbe Interactions: Laboratory Techniques*, Springer Protocols Handbooks, https://doi.org/10.1007/978-1-0716-1080-0_38, © Springer Science+Business Media, LLC, part of Springer Nature 2021

5. Determine the maximum chlorophyll fluorescence (F'_m) by providing 2 min of dark readaptation of actinic white light (430 μmol photons m^2 s) switched on and saturating pulse (8000 μmol photons m^2 s) for 60 s on 15 min interval.
6. Determine the lowest fluorescence intensity (F'_o) by brief interruption of actinic illumination in the presence of far red light.
7. Determine the steady state of fluorescence intensity during actinic illumination (F_s).
8. Calculate the quenching of non-photochemical dissipation of absorb light by following formula:
$$\mathrm{NPQ}\Omega(F_m - F'_m)/F'_m.$$
9. Calculate coefficient for photochemical quenching, q_P of PSII reaction centers by given formula:
$$(F'_m - F_s)/(F'_m - F'_o).$$
10. Calculate the quantum efficiency of PSII photochemistry, φPSII:
$$\varphi\mathrm{PSII} = (F'_m - F_s)/F'_m.$$

Reference

1. Maxwell K, Johnson GN (2000) Chlorophyll fluorescence—a practical guide. J Exp Bot 51:659–668

Chapter 39

Determination of Gas Exchange Measurements

Abstract

Measuring gas exchange is the most commonly used technique to measure photosynthesis of single leaf, entire plants, or plant canopy. This method provides direct measure of the net rate of photosynthetic carbon assimilation. The fundamental advantages of gas exchange measurements are nondestructive, direct, and instantaneous.

Keywords Photosynthetic, Carbon assimilation, Measuring gas exchange

1 Principle

This method provides direct measure of the net rate of photosynthetic carbon assimilation. CO_2 exchange systems use enclosure methods, where the leaf in a closed transparent chamber. The rate of CO_2 fixed by the leaf enclosed is determined by measuring the change in the CO_2 concentration of the air flowing across the chamber. The ambient atmospheric CO_2 concentration is only 0.04% (400 ppm), it is difficult to measure photosynthetic CO_2 uptake and sensitive sensors are needed [1].

2 Materials

1. Hasting flow meter.
2. Beckman infrared gas analyzer.
3. CO_2 gas.

3 Methodology

1. Adjust the leaf temperature to 28 °C.
2. Humidify CO_2 free air and pass it through a condenser at 17 °C to maintain 19 m bar vapor pressure.

M. Senthilkumar, N. Amaresan and A. Sankaranarayanan, *Plant-Microbe Interactions: Laboratory Techniques*, Springer Protocols Handbooks, https://doi.org/10.1007/978-1-0716-1080-0_39,

3. Add CO_2 by Hastings flow meter and connect to a Hastings flow controller.
4. CO_2 concentration is measured with a Beckman infrared gas analyzer (315 B) after the air is passed through an ice trap.
5. Measure the CO_2 concentration inside the leaf, the calculations of the gas exchange measurements are calculated by following formula

$$SE = \mu_0\omega_0 - \mu_e\omega_e$$

where E is the rate of transpiration per unit leaf area (mol/m^2 s); μ_e, μ_0 are the molar flows of air entering and leaving the chamber, and %, ω_0 are the mole fractions of water vapor of the incoming and outgoing airstreams respectively.

Reference

1. Von Caemmerer SV, Farquhar GD (1981) Some relationships between the biochemistry of photosynthesis and the gas exchange of leaves. Planta 153:376–387

Chapter 40

Drought Tolerance Efficiency

Abstract

Drought is one of the important stress conditions, which limit the crop production globally. The ability of the plant to maintain its biomass production during drought condition is called drought tolerance capacity. There are various defense mechanisms, e.g., maintenance of membrane stability, hormone regulation, carbon fixation rate, induction of stress proteins, and generation of antioxidants that play vital role in plants to survive in drought condition.

Keywords Drought stress, Crop productivity, Membrane stability, Crop production

1 Principle

Water is the main growth factor of the plant. In the absence of water, survival of plant is difficult. Plants which are grown in drought area have adaptability to induce drought tolerance mechanism and have induced system as well as growth promotion ability [1, 2].

2 Methodology

1. Collect specific sessional seeds of crops to check drought tolerance efficiency (DTE).
2. Arrange each pot in completely randomized factorial design under two different stress conditions.
3. To evaluate the drought effect, grow the seeds in non-stressed condition and stressed condition.
4. Provide additional fertilizers such as nitrogen and phosphorus to the seeds prior to use for experiment.
5. Record the plant growth on regular interval and harvest the plant after specific time period.
6. Determine crop yield under stressed and non-stressed conditions.

M. Senthilkumar, N. Amaresan and A. Sankaranarayanan, *Plant-Microbe Interactions: Laboratory Techniques*, Springer Protocols Handbooks, https://doi.org/10.1007/978-1-0716-1080-0_40,

7. Calculate the drought tolerance efficiency using following formula

$$\text{DTE} = \text{Yield under stress/Yield under non stress} \times 100$$

Also calculate drought susceptibility index (DSI)

$$\text{DSI} = [1 - (\Upsilon_s/\Upsilon_p)]/\left[1 - \left(\hat{\Upsilon}_s/\hat{\Upsilon}_p\right)\right]$$

where,

Υ_s = Yield of crop with stress.
Υ_p = Yield of crop without stress.
$\hat{\Upsilon}_s$ = mean yield of all cultivars under stressed condition.
$\hat{\Upsilon}_p$ = mean yield of all cultivars under non-stressed condition.

3 Note

- Perform the experiment in multiple replicates for both the conditions.

References

1. Fischer KS, Wood G (1981) Breeding and selection for drought tolerance in tropical maize. In: Proc. symp. on principles and methods in crop improvement for drought resistance with emphasis on rice, IRRI, Philippines
2. Yücel D, Mart D (2014) Drought tolerance in chickpea genotypes. Turk J Agricult Nat Sci 1:1299–1303

Chapter 41

Estimation of Total Phenol Content of Plant Tissues

Abstract

The nonstructural phenolic compounds perform a different activity for example acting as an antioxidant. The Folin–Ciocalteu assay is used for the estimation of phenol content of plant tissue. The Folin–Ciocalteu assay is based on the alteration of electrons in alkaline medium from phenolic compounds to phosphomolybdic/phosphotungstic acid complexes.

Keywords Phenol, Folin–Ciocalteu assay, Phenolic compounds, Phosphomolybdic acid

1 Principle

Phenols react with phosphomolybdic acid in Folin–Ciocalteau reagent in alkaline medium and produce blue colored complex (molybdenum blue) which is measured at 650 nm spectrophotometrically [1].

2 Materials

- Ethanol (80%: 80 mL in 20 mL of distilled water).
- Folin–Ciocalteau reagent.
- Sodium carbonate.
- Mortar and pestle.
- Pipettes.
- Spectrophotometer.

3 Methodology

1. Harvest 0.5–1.0 g of leaf sample and homogenize it in a sterile mortar and pestle using 10 times the volume of 80% ethanol.

M. Senthilkumar, N. Amaresan and A. Sankaranarayanan, *Plant-Microbe Interactions: Laboratory Techniques*, Springer Protocols Handbooks, https://doi.org/10.1007/978-1-0716-1080-0_41,

2. Centrifuge the homogenate at 11,180 × *g* for 20 min. Collect the supernatant with five times the volume of 80% ethanol and centrifuge the pool of the supernatant.
3. Evaporate the supernatant to dryness.
4. Dissolve the residue in a known volume of distilled water.
5. Pipette out different aliquots (0.2–2 mL) into the test tube.
6. Make up the volume in each tube to 3 mL with distilled water.
7. Add 0.5 mL of Folin–Ciocalteau reagent.
8. After 3 min, add 2 mL of 20% Na_2CO_3 solution to each tube.
9. Mix thoroughly. Place the tubes in a boiling water bath for exactly 1 min and cool the tubes.
10. Measure the absorbance at 650 nm against the reagent as blank.

Calculation: From the standard curve find out the concentration of phenols in the test sample and express as mg phenols/ 100 g material.

4 Note

- If any white precipitate is observed during boiling, the color may be developed at room temperature for 60 min.
- Express the results in terms of catechol or any other phenol equivalents used as standard.

Reference

1. Mallick CP, Singh MB (1980) Plant enzymology and histoenzymology. Kalyani Publishers, New Delhi, p 286

Chapter 42

Measurement of Electrolyte Leakage

Abstract

Electrolyte leakage (EL) technique is a promising method for the determination or predicting the physiological status of seedling. The high electrolyte leakage (EL) value at longer time period constitutes a loss of dormancy and increase in cellular damage through the year.

Keywords Plant, Cell death, Stress, Electrolyte leakage (EL)

1 Principle

Measurement of electrolyte leakage depicts the stress tolerance in plant. It is a hallmark of intact response in intact plant cells. To check the stress induced injury in plants electrolyte leakage is measured [1].

2 Materials

- Conductivity meter.
- Deionized water.
- Test tubes.

3 Methodology

1. Collect the leaf blade of plants for the measurement of electrolyte leakage.
2. Cute the leaf blade into segments (ca. 1 cm).
3. Keep the segment from the same leaf into 20 mL of deionized water in the test tube.
4. Wash the leaf segments slowly using rotary shaker (100 rpm) at room temperature to remove solute generate from leaf surface and due to cell damage.

M. Senthilkumar, N. Amaresan and A. Sankaranarayanan, *Plant-Microbe Interactions: Laboratory Techniques*, Springer Protocols Handbooks, https://doi.org/10.1007/978-1-0716-1080-0_42,

5. Measure electrolyte leakage at various washing intervals like (0, 15, 30, 45, 60, 75, and 90 min) using a LF 92 conductivity meter.
6. Check electrolyte leakage during 15 min interval.
7. Consider only newly released electrolyte in an account and measure using following formula.
   Example: electrolyte leakage after 15 min washing period
   $(EC_{15} - EC_0)$/leaf fresh weight
   where, EC_0 represent conductivity at beginning
   EC_{15} represent conductivity after 15 min respectively.

Reference

1. Bajji M, Kinet JM, Lutts S (2002) The use of the electrolyte leakage method for assessing cell membrane stability as a water stress tolerance test in durum wheat. Plant Growth Regul 36:61–70

Check for updates

Chapter 43

Determination of Physiological Traits Related to Abiotic Stresses

Abstract

The abiotic stress is turning into a severe problem mainly in arid and semiarid regions. To survive in the stress condition various changes occur in plants: numerous morphological, physiological, and biochemical changes. For identification of stress resistance cultivars, electrolyte leakage and osmotic potential (OP) has been recommended. The relative water content, high chlorophyll "b," and soluble sugar are related to resistance. Plant RWC is an important tool for measuring the plant water potential (PWP) and turgor potential. Leaf water potential (LWP) involves in avoidance of dehydration mechanism. This protocol provides different physiological traits with their determination methods.

Keywords Abiotic, Physiological traits, PWP, RWC, OP, LWP, Soluble sugar

1 Relative Water Content (RWC)

1.1 Principle

Relative water content is a useful indicator of the state of water balance of a plant essentially because it expresses the absolute amount of water, which the plant requires to reach artificial full saturation. Thus, there are a relationship between RWC and water potential. Turgidity is the function of available water and responsible for deciding physiological processes occurring in leaf tissue [1].

1.2 Materials

- Hot plate/microwave/hot air oven.
- Water bath.

1.3 Methodology-1 [2]

1. Collect 1 g of fresh leaf sample.
2. Dry the leaf sample at 104 °C for 2 h or 72 h at 80 °C.
3. Weigh the dry matter of sample.
4. Calculate the leaf water using following formula:

$$\text{Water content } (\%) = (W_f - W_d)/W_f \times 100$$

where, W_f, fresh weight and W_d, dry weight.

M. Senthilkumar, N. Amaresan and A. Sankaranarayanan, *Plant-Microbe Interactions: Laboratory Techniques*, Springer Protocols Handbooks, https://doi.org/10.1007/978-1-0716-1080-0_43,

1.4 Methodology 2 [3]

1. Take composite sample of leaf disk and weigh fresh weight of leaf sample.
2. Make leaf sample floated on water for up to 4 h.
3. Weigh turgid weight of floated leaf sample after 4 h.
4. Allow the leaf sample to dry at 85 °C and measure dry weight of sample.
5. Calculate relative water content using following formula:

$$\text{Relative water content} = \frac{(\text{Fresh weight}) - (\text{Dry weight})}{(\text{Turgid weight}) - (\text{Dry weight})} \times 100$$

2 Determination of Osmotic Potential

2.1 Principle

Osmotic or solute potential is a main component of water potential which reflects the amount of solutes dissolved in plant tissues. The addition of solute particles to a solvent (water in plants) changes the free energy of the solvent molecules, which allow us indirect means (vapor pressure, freezing point, or boiling point) for the measurement of osmotic potential. The measurement of osmotic potential can be made indirectly by comparing one of the solution's colligative properties with the corresponding cardinal property of the pure water. The osmotic potential can be measured by using osmometer based on the depression of the freezing point or by modern osmometer based on the measurement of vapor pressure depression. The vapor pressure decrease in a solution is directly proportional to the amount of solutes added to it (Raoult's law). The measurement of vapor pressure depression is made by thermocouple hygrometry [4].

2.2 Materials

- Osmometer.
- KCl.
- NaCl.
- Aluminum foil.
- Liquid nitrogen.
- Eppendorf tubes.
- Centrifuge.

2.3 Methodology

1. Harvest the plant samples and chopped into small leaf discs. Cover the leaf discs with aluminum foil and freeze in liquid nitrogen.
2. Thaw the tissue at room temperature for 30 min.

3. For collection of sap, place the leaf discs in eppendorf tubes, cut the bottom of the tube. Place the tube into another fresh tube for collecting the sap.
4. Centrifuge the samples at 11,180 × *g* for 10 min and harvest the sap in the lower tube.
5. Measure the osmotic potential of the sap using vapor pressure osmometer.

2.4 Note

- *Calibration of the instrument*: The osmometer should be calibrated using standard solution 290, 1000, and 100 mmoles/kg.
- *Preparation of calibration curve*: Prepare different concentrations of KCl and NaCl solutions (0.05–0.50) and measure the osmolality at various concentrations. Plot the osmolality values to osmotic potential values to get a calibration curve.

3 Determination of Leaf Water Potential

3.1 Principle

The plants having high leaf water potential (LWP) is involved in avoidance of dehydration mechanism. Hence, plants having high water potential can survive under drought conditions and also associated with drought tolerance indicator like stomal conductance and leaf rolling mechanisms. This can be measured by applying pressure to the detached leaf to return to the water interface. The osmotic potential of the xylem sap is usually less than 0.02 MPa, the hydrostatic pressure in the xylem is equal to the water potential [5].

3.2 Materials

- Pressure chamber apparatus.
- Thermocol ice box.
- Dry nitrogen gas.
- Magnifying hand lens.

3.3 Methodology

1. Harvest the leaf samples and cut their petiole.
2. After excision at petiole, leafs are put into butter paper bags. All such bags are enclosed in a polythene envelope.
3. Keep wrapped envelopes in thermocol ice box containing ice cubes.
4. Insert leaf sample in a pressure chamber apparatus with petiole protruding out from the airtight gasket.
5. Pass compressed dry nitrogen gas slowly in constant motion through flow regulator until xylem sap oozes out at the cut end of the petiole.

6. Increase the pressure in the chamber and observe end of the leaf from the side.
7. At every 4–5 s increase the pressure up to one bar.
8. When xylem sap first appears through the cut surface, cutoff the air inlet valve and read the gauge, indicate the pressure required with a negative sign. It is an estimate of xylem water potential.
9. Take enough readings till a constant measurement is obtained.
10. This instrument measures water potential in bars. However, bar may be converted in Pascals as follows:

$$1 \text{ bar} - 10^5 \text{Pa} = 0.1 \text{ MPa}.$$

3.4 Note

- Use magnifying lens to observe side of the end of the petiole.

4 Determination of Specific Leaf Area

4.1 Principle

The specific leaf area (SLA) is the surface of the leaf which gives total capacity of plant respiration and photosynthesis [6].

4.2 Materials

- Leaf area meter.
- Scale.
- Electronic balance.
- Butter paper.
- Paper knife.

4.3 Methodology 1 [7]

1. Collect 50 different leaflets including small leaflets and large leaflets from the plantation.
2. After cutting the leaflets, keep it in the plastic bag and bring to cold room as soon as possible to prevent shrinkage.
3. Keep leaflet on the table by giving appropriate number.
4. Measure the length (L) and middle width (W) in cm using steal scale.
5. Find specific area of leaflet using following formula:

$$\text{SLA} = L \times W \times f$$

where, f = empirical coefficient of leaf.

4.4 Methodology 2 (by Leaf Area Meter)

1. Initially choose the leaflets, after chosen detach from petiole.
2. Arrange all leaflets on the table by giving number.

3. After placing on the table, measure leaflet area by portable leaf area meter.
4. Record the leaf area with maximum width, average width, and maximum length.
5. Compare the recorded maximum length with manually measured length and then adjusts the length of the leaf area meter.

4.5 Methodology 3

1. Collect leaves from the different leaves canopy.
2. Spread each leaves over the millimeter graph paper, and draw the outline of leaf.
3. Measure the area of each leaf using leaf area meter.
4. Using paper knife cut drawn area of butter paper and weigh on an electronic balance.
5. 1 cm^2 of the same millimeter graph paper is also cut and weigh.
6. Calculate leaf area using following equation:

$$\text{Leaf area } (\text{cm}^2) = x/y$$

where x is the weight of the graph paper covered by the leaf outline (g) and y is the weight (g) of the cm^2 area of the graph paper.

4.6 Note

- Before measurement, leaf area meter is to be calibrated to correct the leaf area of each sample.

5 Determination of Leaf Area Dry-Matter Content

5.1 Principle

Leaf dry matter is the oven dry leaf mass and identified its dry mass in the presence of water saturated fresh mass expressed in mg/g [8].

5.2 Materials

- Hot air oven.
- Leaf area meter.
- Cold box.

5.3 Methodology [9]

1. Collect the sample (stem and twig segment bearing leaf) and wrap it in moist paper and conserved in cold box until further processing.
2. Make a cut on the sample and keep the sample in the tube submerged in deionized water and store in dark at 4 °C for 24 h.
3. After rehydration of the youngest leaf, fully expanded leaf, which are free from herbivore or pathogen damage.

4. Measure the saturated fresh mass of leaves.
5. Determine the leaf projected area using area meter device.
6. Determine the leaf thickness using a linear variable displacement transducer (LVDT).
7. Calibrate the leaf thickness of leaves with metal strip of know thickness.
8. Ovendry the leaves at 60 °C for at least 2 days and determine their dry mass.

$$\text{LDMC} = 1/(\rho\text{F} \times \text{SLA} \times L\text{th})$$

*L*th and the average density of the leaf (ρF).

6 Determination of Leaf Area Index

6.1 Principle

The plant leaf is the basic sites of energy and mass exchange. Important metabolic process such as canopy interception, evapotranspiration, and gross photosynthesis are directly proportional to leaf area index LAI. LAI is the total one-sided green leaf area per unit of ground surface, is an important structural property of vegetation [10].

6.2 Materials

- High resolution camera.

6.3 Methodology

1. Take a photograph using high resolution camera of plant from below or above the canopy to measure LAI.
2. Calculate the leaf area index by measuring half of the area of all leaves per unit area of ground.

$$\text{LAI} = \text{Total leaf area/Land area}$$

References

1. Barr HD, Weatherley PE (1962) A re-examination of the relative turgidity technique for estimating water deficit in leaves. Aust J Biol Sci 15:413–428
2. González L, González-Vilar M (2001) Determination of relative water content. In: Reigosa Roger MJ (ed) Handbook of plant ecophysiology techniques. Springer, Dordrecht, pp 207–212
3. Larcher W (1995) Physiological plant ecology. Springer, Berlin
4. Pearcy RW, Schulze ED, Zimmermann R (1989) Measurement of transpiration and leaf conductance. In: Pearcy RW, Ehleringer J, Mooney HA, Rundel PW (eds) Plant physiological ecology: field methods and instrumentation. Chapman and Hall, New York, pp 137–160
5. Scholander PF, Hammel HT, Brandstreet ET, Hemmingsen E (1965) Sap pressure in vascular plants. Science 148:339–346
6. Nageswara Rao RC, Wright GC (1994) Stability of the relationship between specific leaf area

and carbon isotope discrimination across environments in peanut. Crop Sci 34:98–103

7. Pandey SK, Singh H (2011) A simple, cost-effective method for leaf area estimation. J Bot 2011:1–6
8. Vaieretti MV, Díaz S, Vile D, Garnier E (2007) Two measurement methods of leaf dry matter content produce similar results in a broad range of species. Ann Bot 99:955–958
9. Garnier E, Shipley B, Roumet C, Laurent G (2001) A standardized protocol for the determination of specific leaf area and leaf dry matter content. Funct Ecol 15:688–695
10. Watson DJ (1947) Comparative physiological studies on the growth of field crops: I. Variation in net assimilation rate and leaf area between species and varieties and within and between years. Ann Bot 11:41–76

Part IV

Techniques for Studying Biological Control of Plant Pathogens

Check for updates

Chapter 44

Isolation of Bacteria with Biocontrol Activity Against Phytopathogens: Dual Plate Assay

Abstract

Biological control briefly noted as biocontrol has been used in different fields of biology especially in entomology and plant pathology. In plant pathology, the microbial antagonists suppress the diseases. The organism which is used to suppress the pest or pathogen is referred as biological control agents (BCA). If BCA is a microorganism, then it is referred as microbial biological control agents (MBCA). The MBCAs have a wide mode of action against the pathogenic organisms. MBCAs interact with plants by inducing its resistance. They create nutrient competition to pathogens or involving with other mechanisms which disturbed the growth of the pathogens. The stimulation of plants defense mechanisms by MBCA is mainly dependent on the plant genotypes. Plants enhance the defensive capacity to multiple pathogens by beneficial microbes, various mechanisms like reactive oxygen species, phytoalexins, phenolic compounds/pathogenesis-related proteins (PR proteins), and strategies like physical barriers (cell wall/cuticle modifications) by induced plants and secretion of volatile organic compounds. Dual plate assay, a confrontation assay against the plant pathogens is inhibited by antagonistic bacteria.

Keywords Biocontrol, Phytopathogens, Dual plate assay, Antagonistic activity

Dual plate assay is a confrontation assay against the plant pathogens which is inhibited by antagonistic bacteria. In dual plate assay approach, the pathogenic fungal growth is inhibited by bacteria.

1 Principle

The exudates of antagonistic bacteria diffused through the agar medium and inhibited the growth of the pathogenic microbes on the surface of agar plates. Based on the efficiency of the exudates, the degree of inhibition may vary. Moreover, the degree of inhibition is varying from one to other pathogens and likewise antagonistic bacteria too [1].

M. Senthilkumar, N. Amaresan and A. Sankaranarayanan, *Plant-Microbe Interactions: Laboratory Techniques*, Springer Protocols Handbooks, https://doi.org/10.1007/978-1-0716-1080-0_44,

2 Materials

- Potato dextrose agar.
- Conical flask.
- Eppendorf tubes.
- Tryptone soya agar.
- Nichrome wire.
- Test tubes.
- Diseased plant material.
- Pathogenic bacteria.

3 Methods

1. Isolate the fungal plant pathogen/collect the strain from any culture collection center and maintain it.
2. Prepare potato dextrose agar (PDA) and inoculate the culture in it and incubate the plate at room temperature/27 ± 2° C for 5 days. After the growth, store the fungal strain at 4 °C.
3. For antagonistic activity, collect the soil sample from rhizospheric plant and suspend it in sterile in a 250 ml conical flask containing distilled water, then kept in shaker at 120 rpm for 10 min.
4. Serially dilute the mixture upto 10^{-7}, then transfer 100 μl onto the surface of tryptone soya agar plates.
5. Incubate the plates at 35 ± 2 °C for 24–48 h. After growth, check the potential isolate.
6. Obtain the purified culture by streak plate method, maintain in eppendorf tube in tryptone soya broth contain 20% glycerol at −80 °C for further use.
7. For in vitro phytopathogenic activity of bacteria, prepare PDA medium and in the center of the plate, place 5 mm agar disc containing the actively growing culture of fungal strain.
8. Streak the bacterial culture 3 cm away from the fungal disc toward the edge of the petri plate.
9. Maintain control plate without bacteria.
10. Inhibition of mycelia growth toward the direction of the bacterial isolates indicates the antagonistic activity.
11. Calculate the percentage of a radial mycelial growth inhibition by the following formula:

$$\text{Inhibition (\%)} = \frac{(\text{D1} - \text{D2})}{\text{D1}} \times 100$$

where D1—diameter of radial growth of pathogenic fungi in control; D2—diameter of radial growth of pathogenic fungi with antagonistic bacteria.

4 Observations

- Observe the inhibition and take the note of efficiency of inhibition by the antagonistic bacteria against pathogenic fungi.

5 Precautions

- Pouring of more medium in agar and less medium in agar may affect the result. In general, 15–18 ml of sterilized molten agar/3 mm thickness of agar medium in the plate is an ideal one to carry out to study the dual plate assay.

Reference

1. Tariq M, Yasmin S, Hafeez FY (2010) Biological control of potato black scurf by rhizosphere associated bacteria. Braz J Microbiol 41:431–459

Check for updates

Chapter 45

Antagonistic Activity of Volatile Compound of Bacteria Against Phytopathogens: Dual Plate Assay

Abstract

Microbial biological control agents (MBCA) especially bacteria mediate various mechanisms against phytopathogens. Secretion of volatile organic compounds also played a role in biological control activities. Volatile organic compounds (VOC) basically have easy spreading for a long distance and directly inhibited the growth of the fungi and induced the systemic resistance. The antagonistic activity of volatile compounds of bacteria against the phytopathogens by dual plate assay method is explained here.

Keywords Volatile compounds, Antagonistic activity, Phytopathogens, VOC

1 Principle

The diffusability nature of volatile compounds diffuses the pathogenic fungi and attacks the cellular compounds [1–3].

2 Materials

- Potato dextrose agar.
- Conical flask.
- Eppendorf tubes.
- Tryptone soya agar.
- Nichrome wire.
- Test tubes.
- Diseased plant material/pathogenic bacteria.
- Burette, pipettes.

M. Senthilkumar, N. Amaresan and A. Sankaranarayanan, *Plant-Microbe Interactions: Laboratory Techniques*, Springer Protocols Handbooks, https://doi.org/10.1007/978-1-0716-1080-0_45,

3 Methods [4]

1. Isolate the fungal plant pathogen/collect the strain from any culture collection center and maintain it.
2. Prepare potato dextrose agar (PDA) and inoculate the culture in it and incubate the plate at room temperature/27 ± 2 ° C for 5 days. After the growth, store the fungal strain at 4 °C.
3. For antagonistic activity, collect the soil sample from rhizospheric plant and suspend it in sterile in a 250 ml conical flask containing distilled water, then kept in shaker at 120 rpm for 10 min.
4. Serially dilute the mixture upto 10^{-7}, then transfer 100 μl onto the surface of tryptone soya agar (TSA) plates.
5. Incubate the plates at 35 ± 2 ° C for 24–48 h. After growth, check the potential isolate.
6. Obtain the purified culture by streak plate method, maintain in eppendorf tube in tryptone soya broth contain 20% glycerol at −80 ° C for further use.
7. Prepare the plate contains the lawn growth of bacteria on the surface of TSA plate.
8. Replace the lid of petri plate contains antagonistic bacteria with 7 cm agar plate containing active growth of fungal pathogen.
9. The petri plates inoculated with fungal pathogens were inverted over the plates containing antagonistic bacteria.
10. Seal the plates together with parafilm.
11. Maintain control plate without bacteria in the bottom plate.
12. Incubate the petri dish at room temperature/27 ± 2 ° C and observe the plate at 24 h interval for 72 h.
13. Determine the mycelia growth inhibition by the following formula.

$$\text{Inhibition (\%)} = \frac{\text{D1} - \text{D2}}{\text{D}_1} \times 100$$

where D1 is the diameter of radial growth of pathogenic fungi in control; D2 is the diameter of radial growth of pathogenic fungi with antagonistic bacteria.

14. Perform GC–MS analysis to find out the volatile compounds present in antagonistic bacteria.

4 Observations

- Observe the inhibition of the pathogenic fungi by volatile compounds.

References

1. Brooks DS, Gozales CF, Appel DN, Filer TH (1994) Evaluation of endophytic bacteria as potential biological control agents for oak wilt. Biol Control 4:373–381. https://doi.org/10.1006/bcon.1994.1047
2. Cazar ME, Schmeda-Hirschmann G, Astudillo L (2005) Antimicrobial butyrolactone I derivatives from the Ecuadorian soil fungus *Aspergillus terreus* Thorn. var terreus. World J Microbiol Biotechnol 21:1067–1075. https://doi.org/10.1007/s11274-004-8150-5
3. Fokkema NJ (1978) Fungal antagonism in phyllosphere. Ann Appl Biol 89:115–119
4. Lee S, Hung R, Yap M, Bennett JW (2015) Age matters: the effects of volatile organic compounds emitted by *Trichoderma atroviride* on plant growth. Arch Microbiol 197:723–727

Chapter 46

Detection of Pyoluteorin by Thin Layer Chromatography

Abstract

Pyoluteorin, an aromatic phenolic polyketide antibiotic is capable of suppression of broad spectrum of plant pathogens. It was first isolated from *Pseudomonas aeruginosa*. It is capable to act as bactericidal, herbicidal, and fungicidal agent. It acts as an inhibitor of fungal respiratory chains.

Pyoluteorin was first isolated from *Pseudomonas aeruginosa*. Later, Benicini et al. (Soil Biol Biochem 15:491–492, 1983) reported the pyoluteorin from *P. fluorescence* Pf-5 and CHA0 stains. The first report on the fungicidal activity of pyoluteorin is on *Pythium* spp.

Keywords Pyoluteorin, TLC, Antifungal, Antibiotic, Herbicidal, Bactericidal

Pyoluteorin, an aromatic phenolic polyketide antibiotic, first isolated from *Pseudomonas aeruginosa*. Later Benicini et al. [1] reported the pyoluteorin from *P. fluorescence* Pf-5 and CHA0 stains [1]. The first report on the fungicidal activity of pyoluteorin is on *Pythium* spp.

1 Principle

In the thin layer chromatography technique, stationary (solid phase) and moving (liquid phase) phases are used. In this adsorption process, the solute competes with the solvent on the surface of adsorbent. Based on distribution coefficients, the compounds are distributed on the surface of adsorbent [2].

2 Materials

- Bacterial culture especially fluorescence *Pseudomonas* sp.
- Pigment production medium (peptone, 20 g; glycerol, 20 ml; NaCl 5 g; KNO_3—1 g; distilled water 1 l; pH 7.2).
- 1 N HCl; Ethyl acetate solution (30.67 ml HCl in 1 l of distilled water).

M. Senthilkumar, N. Amaresan and A. Sankaranarayanan, *Plant-Microbe Interactions: Laboratory Techniques*, Springer Protocols Handbooks, https://doi.org/10.1007/978-1-0716-1080-0_46, © Springer Science+Business Media, LLC, part of Springer Nature 2021

- TLC coated plate with 250 μm silica gel.
- Chloroform, methanol, benzene and acetic acid (TLC grade).

3 Methods

1. Grow the bacterial culture in 5 ml pigment production medium for 4 days at rotary shaker at room temperature.
2. Centrifuge the culture at 2000 × *g* for 5 min.
3. Collect the supernatant and acidified with pH 2 in 1 N HCl.
4. Extracted with equal volume ethyl acetate solution.
5. Dry the extract and the residue dissolved in methanol.
6. Pyoluteorin developed in chloroform and methanol (9:1 v/v and benzene and acetic acid (9.5:0.5 v/v).
7. 20 μl of sample applied in TLC plate.
8. Visualize the spot by spraying with diazotized sulfanilic acid or under UV at 254 nm.
9. Rf value of the spot compared with the standard.

4 Observations

- Observe the spot in TLC plate after sprayed with diazotized sulfanilic acid.

5 Precautions

- Volume of the sample should be minimal and place the spot in TLC plate at a particular point and it should not be spread on TLC plate.
- Caution with the thickness of silica gel and smoothness of the surface of silica gel on TLC plate.

References

1. Bencini DA, Howell CR, Wild JR (1983) Production of phenolic metabolites by a soil pseudomonad. Soil Biol Biochem 15:491–492
2. Vinay JU, Naik MK, Rangeshwaran R, Chennappa G, Shaikh SS, Sayyed RZ (2016) Detection of antimicrobial traits in fluorescent pseudomonads and molecular characterization of an antibiotic pyoluteorin. 3 Biotech 6:227

Check for updates

Chapter 47

Detection of Siderophore Producing Microorganisms

Abstract

Universal CAS-Agar Plate Assay for Detection of Siderophore Producing Microorganisms: Siderophores are low-molecular-weight (<1000 Da), Fe (III)-specific ligands that can chelate iron compounds with high affinity. Siderophores are secondary metabolites assembled by non-ribosomal cytoplasmic peptide synthases. Functional groups and lateral chains of siderophore confer a strong affinity toward ferric ion. The affinity of siderophores ranges from low (aerobactin) to high (enterobactin). These compounds are typically produced by various bacteria, fungi, and monocotyledonous plants in response to iron stress. Generally, iron has vital functions in photosynthesis, enzyme cofactor, redox reagent, respiration, synthesis of nucleosides, and amino acids of the living system. Apart from the above functions, siderophores also act as plant growth promoters, biocontrol agents, and bioremediation, in addition to soil mineral weathering. In spite of the concentration of Fe, iron deficiency is induced in plants grown under alkaline soils. Affinity of the plant for Fe is decreased in calcareous soils and impedes iron uptake mechanism.

Based on the coordinating groups that chelate the Fe (III) ion, siderophores are generally classified into catecholates, hydroxamates, and carboxylates. Along with this, other types of siderophores have chemically distinct Fe (III) ion binding group and mixed ligands having coordinating groups that fall into chemically different classes. These groups are usually identified based on their electrophoretic mobility, spectrophotometric titration, proton nuclear magnetic resonance spectroscopy, mass spectrometry, acid hydrolysis, and biological activity.

Modified CAS-Agar Plate Assay for Detection of Siderophore Producing Fungi and Bacteria: The concentration of HDTMA and CAS is crucial for siderophore medium preparation. The lower concentration leads to precipitation of the blue dye and higher concentration may turn toxic to all organisms. To avoid this, universal CAS-agar plate assay (Schwyn and Neilands, Anal Biochem 160:47–56, 1987) was modified (Milagres et al., J Microbiol Methods 37:1–6, 1999) to test siderophore producing ability of several fungi and gram-positive bacteria.

O-CAS Assay for Fast Universal Method to Detect Siderophore Producing Microorganisms: O-CAS means "Overlaid—CAS" assay used for the detection of siderophore producing microorganisms from any samples. This method is used for the detection of more than one siderophore producing microorganism. By adjustment of medium composition, detection is optimized. In the existing CAS universal assay method, the medium not only has the required nutrients, some microorganisms (especially gram-positive bacteria and fungi) are unable to grow due to the presence of inhibitory substances.

Layer Plate CAS Assay for Determining Fungal Siderophores: Hexadecyl tri-methyl ammonium bromide (HDTMA) in CAS assay inhibits the growth of many fungi. N-dodecyl-N, N-dimethyl-3-ammonio-1-propanesulfonate (DDAPS) is less toxic to fungi compared to HDTMA and produces results similar to CAS assay with HDTMA. Hence, DDAPS replaces HDTMA as surfactant in CAS medium. Petri plates having AY + CAS assay solution overlaid with AY agar medium are used for better growth of yeast and fungi.

Keywords Siderophore, FE (III), CAS agar, Cheltors, CAS/HDTMA, Modified CAS agar, O-CAS assay, Siderophore types, Layer plate, Fungal siderophores

M. Senthilkumar, N. Amaresan and A. Sankaranarayanan, *Plant-Microbe Interactions: Laboratory Techniques*, Springer Protocols Handbooks, https://doi.org/10.1007/978-1-0716-1080-0_47, © Springer Science+Business Media, LLC, part of Springer Nature 2021

1 Universal CAS-Agar Plate Assay for Detection of Siderophore Producing Microorganisms

The simplified step-by-step universal CAS-Agar plate method was developed [1] from the originally developed protocol [2] using chrome azurol S (CAS) and hexadecyl-trimethyl-ammonium bromide (HDTMA) as indicators.

1.1 Principle

The CAS/HDTMA tightly complexes with ferric iron to produce a blue color at pH 6.8. When a strong iron chelator such as a siderophore removes iron from the dye complex, the color changes from blue to orange [3].

1.2 Materials

1. *Solution-1*: Dissolve 0.06 g of CAS in 50 ml of ddH_2O.
2. *Solution-2*: Dissolve 0.0027 g of $FeCl_3$-$6H_2O$ in 10 ml of 10 mM HCl (how to prepare 10 mM HCl).
3. *Solution-3*: Dissolve 0.073 g of HDTMA in 40 ml of ddH_2O.
4. *Minimal Medium 9 (MM9) salt solution stock*: Dissolve 15 g KH_2PO_4, 25 g NaCl, and 50 g NH4Cl in 500 ml of ddH_2O.
5. *20% Glucose stock*: Dissolve 20 g glucose in 100 ml of ddH_2O.-Solution should be filter sterilized.
6. *Casamino acid solution*: Dissolve 3 g casamino acid in 27 ml of ddH_2O. Extract with 3% 8-hydroxyquinoline in chloroform to remove any trace iron. Solution should be filter sterilized.
7. *Blue dye*: Mix solution-1 (50 ml) with 9 ml of solution-2. Then mix with solution-3 (40 ml). Autoclave and store it in a plastic container/bottle.
8. Agar (15 g).

1.3 Method

1. Add 100 ml of MM9 salt solution to 750 ml of ddH_2O.
2. Dissolve 32.24 g piperazine-N,N′-bis(2-ethanesulfonic acid) PIPES. PIPES will not dissolve below pH of 5. Bring pH up to 6 and slowly add PIPES while stirring. The pH will drop as PIPES dissolves. While stirring, slowly bring the pH up to 6.8. Don't exceed pH 6.8 as this will turn the solution green.
3. Add 15 g Agar, autoclave it, and cool to 50 °C.
4. Add 30 ml of sterile casamino acid solution and 10 ml of sterile 20% glucose to MM9–PIPES mixture.
5. Slowly add 100 ml of blue dye solution along the glass wall with enough agitation to mix thoroughly.
6. Aseptically pour into petri plates and allow it for solidification.

1.4 Observation

- Determine the number of days required for microorganisms to cover halves of nutrient media.

- Determine the CAS reaction rate by measuring advance of the color change front in CAS-Agar halves.
- CAS-Agar color can be changed from blue to orange, purple, or dark purplish-red.

1.5 Precautions

- Use double distilled water to prepare growth medium.
- All glasswares should be washed with 6 M HCl and rinsed with ddH_2O to remove iron contamination.

2 Modified CAS-Agar Plate Assay for Detection of Siderophore Producing Fungi and Bacteria

The concentration of HDTMA and CAS is crucial for siderophore medium preparation. The lower concentration leads to precipitation of the blue dye and higher concentration may turn into toxic to all organisms. To avoid this, universal CAS-Agar plate assay [2] was modified [4] to test siderophore producing ability of several fungi and gram positive bacteria [5].

2.1 Methodology

1. Prepare petri plates containing suitable agar-based growth medium for specific microorganism.
2. After becoming solidified, one half of the medium from petri plate can be cut, removed, and replaced by CAS-Agar medium. CAS-Agar medium can be prepared as described earlier.
3. The halves containing culture medium should be inoculated with microorganisms as far as possible from the borderline between two media.
4. Inoculated petri plates should be incubated at appropriate condition under dark.
5. One set of petri plates with halves containing growth medium and CAS-Agar without inoculation should be used as control.

2.2 Observation

- Determine the number of days required for microorganisms to cover halves of nutrient medium.
- Determine the CAS reaction rate by measuring advance of the color change front in CAS-Agar halves.
- CAS-Agar color can be changed from blue to orange, purple, or dark purplish-red.

2.3 Precautions

- Use double distilled water to prepare growth medium.
- All glasswares should be washed with 6 M HCl and rinsed with ddH_2O to remove iron contamination.

- Iron content of culture medium should be lower than 1.8 μM as determined by atomic absorption spectrometry.

3 O-CAS Assay for Fast Universal Method to Detect Siderophore Producing Microorganisms

3.1 Materials

Chemicals/solutions/buffers required as per Chap. 47.1 Agar (1.5%) in CAS-Agar is replaced by agarose (0.9% w/v) as gelling agents.

Appropriate growth media for culturing specific microorganisms.

3.2 Method

Prepare petri plates containing suitable agar-based growth medium for specific microorganism, inoculate and incubate at appropriate condition.

CAS-Agarose solution should be applied over agar plates containing cultivated microorganisms.

3.3 Observation

After a maximum period of 15 min, a change in color will be observed in the overlaid medium, exclusively surrounding producer microorganisms from blue to purple (catechol), orange (hydroxamates).

4 Layer Plate CAS Assay for Determining Fungal Siderophores

Hexadecyl-trimethyl-ammonium bromide (HDTMA) in CAS assay inhibits the growth of many fungi. N-dodecyl-N, N-dimethyl-3-ammonio-1-propanesulfonate (DDAPS) is less toxic to fungi compared to HDTMA and produces results similar to CAS assay with HDTMA. Hence, DDAPS replaces HDTMA as surfactant in CAS medium. Petri plate having AY + CAS assay solution overlaid with AY agar medium are used for better growth of yeast and fungi.

4.1 Principle

- As described earlier.

4.2 Materials

- *Acetate–Yeast (AY) medium*: Dissolve sodium acetate (0.27 g), yeast extract (0.15 g), agar (15 g) in 900 ml deionized water, autoclave for 30 min and cool to 50 °C.
- *Trace element solution (TES)*: Dissolve $CuSO_4.5H_2O$ (10 mg), $ZnSO_4.7H_2O$ (44 mg), $CoCl_2.6H_2O$ (20 mg) $Na_2MoO_4.2H_2O$ (13 mg) in 1000 ml of water.
- *10× CAS assay solution*: 10 mM $FeNO_3$ (20 ml), 10 mM chrome azurol S (40 ml), and 10 mM DDAPS (100 ml).

4.3 Methods

1. Prepare 900 ml of AY medium, autoclave it, and then cool to 60 °C.
2. Add 1 ml of TES and 100 ml of CAS assay solution into AY medium, pour into petri plates, and solidify it to form dark blue bottom layer.
3. Pour AY medium over AY + CAS plates as thin layer, solidify, and store it at 4 °C that facilitate diminishing blue gradient from top to bottom.
4. Inoculate active fungal cultures on AY + CAS plates and incubate at 25 °C in the dark for 12 days.

4.4 Observation

- The AYCAS-L plate assay starts off blue and turns yellow as siderophores remove the Fe (III) from AYCAS-L plates.

4.5 Precautions

- Use double distilled water to prepare growth medium.
- All glasswares should be washed with 6 M HCl and rinsed with ddH_2O to remove iron contamination.
- Iron content of culture medium should be lower than 1.8 μM as determined by atomic absorption spectrometry.

References

1. Louden BC, Haarmann D, Lynne AM (2011) Use of blue agar CAS assay for siderophore detection. J Microbiol Biol Educ 12(1):51–53
2. Schwyn B, Neilands JB (1987) Universal chemical assay for the detection and determination of siderophores. Anal Biochem 160:47–56
3. Andrews MY, Santelli CM, Duckworth OW (2015) Layer plate CAS assay for the quantitation of siderophore production and determination of exdutation patterns for fungi. J Microbiol Methods 121:41–43
4. Milagres AM, Machuca A, Napoleao D (1999) Detection of siderophores production from several fungi and bacteria by a modification of chrome azurol S (CAS) agar plate assay. J Microbiol Methods 37:1–6
5. Hu QP, Xu JG (2011) A simple double-layered chrome azurol S agar (SDCASA) plate assay to optimize the production of siderophores by a potential biocontrol agent *Bacillus*. Afr J Microbiol Res 5:4321–4327

Chapter 48

Quantitative Estimation of Siderophore Production by Microorganisms

Abstract

Traditional Method: Microorganisms are in need of iron, the available free metal iron (Fe III) in insoluble form which leads to the minimal availability of iron to microbes. Hence the microbes especially bacteria synthesize low-molecular-weight chelators which form complex with irons. The siderophore form complex with ferric form of iron and enter inside the microbial cell by different mechanisms. After entering inside the cell, the ferric iron reduces into ferrous and is free from siderophore. The quantitative estimation of siderophore using spectrophotometer is explained in this protocol.

Quantitative Estimation of Siderophore Production by Microplate Method: Microplate method is an efficient and rapid method for the quantitative estimation of siderophore. In this method, the quantity of siderophore by several siderophore producing microbes in a single plate is estimated. The other advantages of this method are saving time, chemical, and labor. The supernatant of bacterial culture is added in the well of microplate containing CAS reagent and after incubation, the OD value measures at 630 nm using microplate reader.

Quantitative Estimation of Fungal Siderophore Production by Layer Plate CAS Method: Chrome Azurol S agar assay is used to detect the chelating activity of siderophore, however due to toxicity (HDTMA) the traditional procedure is limited especially for fungi (Alexander and Zuberer, Biol Fertil Soils 12:39–45, 1991). Instead HDTMA, N-dodecyl-N-N-dimethyl-3- ammonio-1-propanesulfonate (DDAPS) along with CAS is used in the modified procedure.

Keywords Siderophore, Estimation, Spectrophotometer, Traditional method, Microplate method, Quantitative estimation, Rapid method, Fungi, CAS, Layer plate method

1 Traditional Method

1.1 Materials

1. *Solution 1*: 3 mol/l hydrochloric acid solution (To make 3 M solution of hydrochloric acid, add 246.351 ml to 250 ml deionized water and adjust the final volume to 1000 ml deionized water).
2. *Solution 2*: CAS reagent. 121 mg of CAS was dissolved in 100 ml of distilled water and 20 ml of 1 mM ferric chloride ($FeCl_3.6H_2O$). This solution was added to 20 ml HDTMA solution by continuous stirring. For HDTMA solution preparation, refer protocol-21. The CAS HDTMA solution was sterilized before further use.

M. Senthilkumar, N. Amaresan and A. Sankaranarayanan, *Plant-Microbe Interactions: Laboratory Techniques*, Springer Protocols Handbooks, https://doi.org/10.1007/978-1-0716-1080-0_48,

1.2 Method

1. Add 1 ml LB broth into 1.5 ml sterilized centrifuge tube, then add 10 μl of freshly grown bacterial culture (10^8CFU/ml). Maintain the replicates and control tube.
2. Incubate the tubes at 28 °C for 48 h.
3. Centrifuge the tubes at 16,800 g for 10 min and collect the supernatant and use the supernatant for estimation of siderophore.
4. 0.5 ml of supernatant mixed 0.5 ml CAS reagent, after 20 min take optical density at 630 nm in spectrophotometer.
5. Measure the siderophore production in percent siderophore unit (psu) by the following formula

$$\text{Siderophore production (psu)} : \frac{(\text{Ar} - \text{As}) \times 100}{\text{Ar}}$$

Ar—Absorbance of reference (CAS solution and uninoculated broth).

As—Absorbance of sample (CAS solution and supernatant).

1.3 Precautions

1. Before starting the experiment, wash the glasswares with 3 mol/l hydrochloric acid solution to remove iron. Then, wash all the glasswares with deionized water.

2 Quantitative Estimation of Siderophore Production by Microplate Method

2.1 Principle

1. As described earlier [Chap. 47].

2.2 Materials

1. As described earlier.

2.3 Method [1]

1. 5 μl of freshly grown bacterial culture (10^8CFU/ml) into 0.5 ml LB broth in a micro-centrifuge tube. Maintain the replicates and control tube.
2. Incubate the tubes at 28 °C for 48 h.
3. Centrifuge the tubes at 10,000 rpm for 10 min and collect the supernatant (100 μl) and add it in the well of micro plate of each bacterial culture.
4. Then add 100 μl of CAS reagent.
5. After the period of incubation, take optical density at 630 nm in spectrophotometer using microplate reader.
6. Replicates and control also maintained.
7. Estimate the siderophore production using the formula as mentioned in [Chap. 48.1].

3 Quantitative Estimation of Fungal Siderophore Production by Layer Plate CAS Method

Chrome Azurol S agar assay used to detect the chelating activity of siderophore, however due to toxicity (HDTMA) the traditional procedure is limited especially for fungi [2]. Instead of HDTMA, N-dodecyl-N-N-dimethyl-3-ammonio-1-propanesulfonate (DDAPS) along with CAS is used in the modified procedure [3].

3.1 Principle

1. As described earlier [Chap. 47].

3.2 Materials

1. N-dodecyl-N-N-dimethyl-3-ammonio-1-propanesulfonate (DDAPS).
2. *Acetate yeast (AY) agar media*: 0.27 g Na-acetate, 0.15 g yeast extract, 15 g agar in 900 ml deionized water. Autoclave it for 30 min, cool to 60 °C, add 1 ml filter sterilized trace element solution (10 mg $CuSO_4.5H_2O$, 44 mg $ZnSO_4$. $7H_2O$, 20 mg $CoCl_2$. $6H_2O$ and 13 mg Na_2Mo_4. $2H_2O$ per liter) with 100 ml of separately autoclaved 10× CAS assay.
3. *CAS assay with DDAPS*: 20 ml of 10 mM $Fe(NO_3)_3$, 40 ml of 10 mM chrome azuros S and 100 ml of 10 mM DDAPS.

3.3 Method

1. Prepare the plate of CAS with DDAPS (CAS + DDAPS) forming a dark blue bottom layer in petri plate (approximately 20 ml).
2. After it cool, then pour the top layer containing acetate yeast agar medium.
3. Store the plate at 4 °C for overnight for diffusion. During the storage period, the CAS assay diffused upward and makes light blue from top to bottom of the plate.
4. Stab a small quantity of mycelium into the center of the plate and then incubate the plate at 25 °C for 12 days in the dark room. Maintain duplicate also.
5. Growth of fungi and color change is noted during the incubation period.
6. To quantify the siderophore, duplicate plate is photographed on a light box and the image is resized into 9 cm. in Adobe Photoshop elements J Microvision 1.2.7.
7. Multiply the measured area by the known amount of CAS assay present per unit area of plate provided the production of siderophore.

3.4 Observation

1. Observe the color change from blue to yellow and note the growth of the fungi (vertical/diffused growth on CAS assay plate).

References

1. Arora NK, Verma M (2017) Modified microplate method for rapid and efficient estimation siderophore produced by bacteria. 3 Biotech 7:381, 1–11
2. Alexander DB, Zuberer DA (1991) Use of chrome azurol S reagents to evaluate siderophore production by rhizosphere bacteria. Biol Fertil Soils 12:39–45
3. Andrews MY, Santelli CM, Duckworth OW (2015) Layer plate CAS assay for the quantitation of siderophore production and determination of exudation patterns for fungi. J Microbiol Methods 121:41–43

Chapter 49

Extraction and Characterization of Siderophore [Catecholate Type] by Thin Layer Chromatography

Abstract

The estimation of catecholate type of siderophore was performed (Jadhav and Desai, Curr Microbiol 24:137–141, 1992), following the protocol originally developed (Arnow, J Biol Chem 118:531–535, 1937) by confirming the presence of 2,3–dihydroxy benzoic acid (2,3-DHBA) and 3,4–dihydroxy benzoic acid (3,4-DHBA). The presence of dihydoxy benzoic acid in the siderophore was detected by TLC method. Further the catechol type of siderophore conjugated with amino acids and detects the amino acid by paper chromatography.

Keywords Siderophore, Catecholate type, TLC, Extraction, Characterization

1 Principle

1. Dihydroxy benzoic acid production in the growth medium depends on the iron level or starved condition of iron in the medium. If the growth medium contains the iron, no dihydoxy benzoic acid was produced. Siderophore has high affinity with Fe^{3+} and forms a stable complex, hence the addition of $FeCl_3$ with culture supernatant yielding wine color confirms the presence of siderophore [1, 2].

2 Materials

1. Yeast *Extract Mannitol agar.*
2. *Siderophore production medium*: 10 g mannitol; 2 g sodium glutamate, 0.5 g KH_2PO_4, 0.2 g $MgSO_4.7H_2O$, 0.1 g sodium chloride. Use double distilled water.
3. *1 mM $FeCl_3$ solution*
4. *Hathway's reagent*: K-ferricyanide 0.3 g, $FeCl_3$ 0.3 g, double distilled water 100 ml.
5. *Ethyl acetate solution.*

M. Senthilkumar, N. Amaresan and A. Sankaranarayanan, *Plant-Microbe Interactions: Laboratory Techniques*, Springer Protocols Handbooks, https://doi.org/10.1007/978-1-0716-1080-0_49,

6. *Solvents*: benzene, toluene, acetic acid, and butanol.
7. *6 N HCl and 1 N NaOH solution*: For making 1 l of HCl, 492.703 ml con. Hydrochloric acid with 250 ml of deionized water and adjust the final volume to 1000 ml with deionized water. Dissolve 40 g of NaOH in 1 l of deionized water.
8. *Ninhydrin reagent*: Ninhydrin (0.2% wt/vol) in acetone.

3 Methods

1. Isolate the *Rhizobium* from the plant and maintain the culture in YEMA (yeast extract mannitol agar with Congo red).
2. Prepare siderophore production medium in double distilled water and inoculate the organism and keep it 22–24 h.
3. Addition of 1 mM $FeCl_3$ solution to the supernatant turning into wine color ensures the presence of catechol type of siderophore.
4. Further follow [2] and confirm the absence of hydroxamic siderophore.
5. For the estimation of catecholate type of siderophore, acidify the culture supernatant with pH 2, then extract with equal volume of ethyl acetate, later evaporate the ethyl acetate.
6. For characterization, TLC with a solvent system containing benzene: toluene: acetic acid (2:2:1), then add Hathway's reagent; ensure the blue colour spot for 2,3–DHBA and 3,4–DHBA.
7. For preparative TLC, the scraped and eluted 2,3–DHBA and 3,4–DHBA is analyzed by UV-spectrophotometric scanning. The spectra matched with authentic 2,3–DHBA and 3,4–DHBA.
8. The acidified siderophore analyzed by paper chromatography with the solvent butanol: acetic acid: water 12:3:5, with the ninhydrin reagent, the chromatogram with ninhydrin reagent and found the presence of alanine and lysine amino acids.

4 Observation

1. Addition $FeCl_3$ solution to the supernatant turns into wine color.
2. Development of blue colour spot in thin layer chromatography.

5 Precautions

1. Use double distilled water to prepare the reagents and medium.
2. All glasswares should be washed with 6 M HCl and rinsed with ddH_2O to remove iron contamination.

References

1. Chakraborty RN, Patel HN, Desai SB (1990) Isolation and partial characterization of catechol-type siderophore from *Pseudomonas stutzeri* RC 7. Curr Microbiol 20:283–286
2. Gibson F, Magrath DI (1969) The isolation and characterization of a hydroxamic acid (aerobactin) formed by Aerobacter aerogenes 62-1. Biochim Biophys Acta 192(2):175–184
3. Jadhav RS, Desai AJ (1992) Isolation and characterization of siderophore from cowpea *Rhizobium* (Peanut isolate). Curr Microbiol 24:137–141
4. Arnow LE (1937) Colorimetric estimation of the components of 3,4-dihydroxy phenylalanine tyrosine mixtures. J Biol Chem 118:531–535

Chapter 50

Purification of Siderophores by High Performance Liquid Chromatography and Electrospray Ionization-Mass Spectrometry (ESI-MS)

Abstract

Many bacteria secrete high affinity and low molecular weight iron scavenging siderophore under iron deficit environment. It is an essential need to purify the siderophore for further characterization, structural elucidation, and acquire more information about it. In the given protocol, the purification of siderophore by high performance liquid chromatography (HPLC) and electrospray ionization-mass spectrometry (ESI-MS) is briefed.

Keywords Siderophore, HPLC, ESI-MS, Characterization

1 Principle

Partition coefficient or distribution coefficient is in which a compound distributes between two immiscible phases like solid/liquid or gas/liquid. Molecular size, shape, ionic nature, and topography of molecules played a role in the distribution inside the column [1, 2].

2 Materials

1. CAS reagent.
2. *Solvents*: Methanol and ethanol.
3. *Sephadex LH-20 column*.
4. Luna C18 column.

M. Senthilkumar, N. Amaresan and A. Sankaranarayanan, *Plant-Microbe Interactions: Laboratory Techniques*, Springer Protocols Handbooks, https://doi.org/10.1007/978-1-0716-1080-0_50,

3 Methods

1. Centrifuge the overnight bacterial broth culture, methanol mixed with the supernatant with 3:1 volume ratio, then add four volumes of ethanol and left overnight at 4 °C.
2. Remove the precipitate, the crude siderophore obtained by vacuum rotary evaporator at 60 °C, purified with a Sephadex LH-20 column. Wash with deionized water and collect the fractions (5 ml) with a help of fraction collector and check the presence of siderophore using CAS reagent. Fractions pooled, analyzed and lyophilized by reverse phase HPLC.
3. Conditions of HPLC as follows: sample volume 20 μl; methanol: water 70:30 v/v as mobile phase, flow rate 1.0 ml/min 30 °C. column temperature; 220 nm detection wavelength.
4. Collect the siderophore corresponding to the peak, dehydrate via lyophilization and analyzed by electrospray ionization-mass spectrometry (ESI-MS).
5. Settings: 90 °C source temperature; 250 °C desolvation temperature; capillary voltage 4500 V; flow rate 6 l/min; 200–800 mass range. Collided the ion of interest with helium gas to get fragments and analyzed based *m/z*.
6. Peak of HPLC chromatogram analyzed by ESI-MS.
7. Analyze the peak of ESI-MS and calculate the molecular weight of siderophore produced by the bacteria.

4 Observation

1. Observe the sharp peak both at HPLC chromatogram and ESI-MS spectrum.

References

1. Yu S, Teng C, Liang J, Song T, Dong L, Bai X, Jin Y, Qu J (2017) Characterization of siderophore produced by *Pseudomonas syringae* BAF 1 and its inhibitory effects on spore germination and mycelium morphology of *Fusarium oxysporum*. J Microbiol 55(11):877–884
2. Sajdowich S, Haller JC, Krafft AE, Hunsucker SW, Mant CT, Duncan MW, Hodges RS, Jones DNM, Holmes RK (2012) Purification and structural characterization of siderophore (Corynebactin) from *Corynebacterium diptheriae*. PLoS One 7(4):e34591. https://doi.org/10.1371/journal.pone.0034591

Chapter 51

Isolation of Microorganisms Producing Hydrolytic Enzymes: Protease

Abstract

Proteins can be hydrolyzed by the secretion of extracellular protease enzyme. The hydrolytic enzyme, protease, breakdowns the peptide bonds between proteins. Proteolytic bacteria use the enzymes and hydrolyze the casein present in skim milk agar medium. The protocol briefed about the isolation of microbes producing protease enzyme.

Keywords Protease enzyme, Hydrolytic enzyme, Skim milk agar, Hydrolysis

1 Principle

Proteases are enzymes that break the proteins to smaller peptides or amino acids. Proteases are of various types namely serine, cysteine, metallo aspartate and threonine. *Bacillus, Aspergillus, Pseudomonas,* etc. are the organisms that produce protease. The mode of mechanism of this enzyme is used to cleave a peptide bond in the substrate [1–4].

2 Materials

1. Skim milk agar
2. Soil sample.

3 Methods

1. Weigh 1 g of soil sample and dispensed it into sterile water, then serially diluted it.
2. Prepare skim milk agar plates.
3. Transfer the soil sample into sterile petri plates and spread over the surface of agar plates using L-rod (spreader).

M. Senthilkumar, N. Amaresan and A. Sankaranarayanan, *Plant-Microbe Interactions: Laboratory Techniques*, Springer Protocols Handbooks, https://doi.org/10.1007/978-1-0716-1080-0_51,

4. Keep the plates in incubator.
5. After the period of incubation, observe the plates for halo surrounds around the growth of the organisms.

4 Observation

Observe the halo zone around the potential enzyme producing organism in the skim agar plates.

References

1. Ryden A, Lindberg M, Philipson L (1973) Isolation and characterization of two protease-producing mutants from Staphylococcus aureus. J Bacteriol 116(1):25–32
2. Suberu Y, Akande I, Samuel T, Lawal A, Olaniran A (2019) Optimization of protease production in indigenous *Bacillus* species isolated from soil samples in Lagos, Nigeria using response surface methodology. Biocatal Agric Biotechnol 18(2019). https://doi.org/10.1016/j.bcab.2019.01.049
3. Mazzucotelli CA, Ponce AG, Kotlar CE, Moreira MR (2013) Isolation and characterization of bacterial strains with a hydrolytic profile with potential use in bioconversion of agro industrial by-products and waste. Food Sci Technol 33 (2):295–303
4. Kuddus M, Ramteke PW (2011) Production optimization of an extracellular cold-active alkaline protease from *Stenotrophomonas maltophilia* MTCC 7528 and its application in detergent industry. Afr J Biotechnol 5(7):809–816

Chapter 52

Isolation of Microorganisms Producing Hydrolytic Enzymes: Lipase

Abstract

Lipase enzyme producing organism from the soil sample can able to hydrolyze the tributyrin by the secretion of extracellular lipase enzyme. Plant pathogens released cuticle degrading enzymes; the lipophilic cuticle covers the plant leaf surface. Various defense-associated lipases include phospholipases, phospholipase A (PLA), PLA1, and PLA2 involved in the plant defense process against plant pathogens.

Keywords Lipase, Hydrolytic enzymes, Tributyrin agar, Exoenzymes

Lipase enzyme producing organism from the soil sample which can able to hydrolyze the tributyrin by the secretion of extracellular lipase enzyme.

1 Principle

Tributyrin agar is a differential medium that tests the ability of an organism to produce an exoenzyme, called lipase that hydrolyzes tributyrin oil. Lipases break down lipids (fats). Tributyrin oil is a type of lipid called a triglyceride. Lipase allows the organisms that produce it to break down lipids into smaller fragments [1–4].

2 Materials

1. Tributyrin agar.
2. Soil sample.

3 Procedure

1. Weigh 1 g of soil sample and dispensed it into sterile water, then serially diluted it.

M. Senthilkumar, N. Amaresan and A. Sankaranarayanan, *Plant-Microbe Interactions: Laboratory Techniques*, Springer Protocols Handbooks, https://doi.org/10.1007/978-1-0716-1080-0_52, © Springer Science+Business Media, LLC, part of Springer Nature 2021

2. Prepare Tributyrin agar plates.
3. Transfer the soil sample into sterile petri plates and spread over the surface of agar plates using L-rod (spreader).
4. Keep the plates in incubator.
5. After the period of incubation, observe the plates for halo surrounds around the growth of the organisms.

4 Observation

Observe the halo zone around the potential enzyme producing organism in the tributyrin agar plates.

References

1. Watanabe N, Ota Y, Minoda Y, Yamada K (1977) Isolation and identification of alkaline lipase producing Microorganisms, cultural conditions and some properties of crude enzymes. Agric Biol Chem 41(8):1353–1358
2. Lee H, Park OK (2019) Lipases associated with plant defense against pathogens. Plant Sci 279:51–58. https://doi.org/10.1016/j.plantsci.2018.07.003
3. Oliveira F, Souza CE, Peclat VORL, Salgado JM, Ribeiro BD, Coelho MAZ, Venancio A, Belo I (2017) Optimization of lipase production by *Aspergillus ibericus* from oil cakes and its application in esterification reactions. Food Bioprod Process 102:268–277
4. Selvakumar P, Sivashanmugam P (2017) Optimization of lipase production from organic solid waste by anaerobic digestion and its application in biodiesel production. Fuel Process Technol 165:1–8

Check for updates

Chapter 53

Isolation of Bacteria Producing Thermo Stable Hydrolytic Enzyme: Chitinase

Abstract

Chitin (1,4-2acetamide 2-deoxy beta-D-Glucan) a polysaccharide contains N-acetyl glucosamine (Glc-NAc) residues linked by beta-1,4 bonding. Chitinase is involved in various pivotal roles including defense, nutrient digestion, and pathogenesis. Chitinase enzyme hydrolyzes the chitin polymers high molecular long chain compounds into short chain low molecular compounds. In the defense mechanism, microbial chitinases weaken the cell wall of many pests and pathogens. For the isolation of chitinase enzyme producing bacteria, chitin containing medium is used.

Keywords Chitinase, Glycosidic linkage, Chitin

1 Principle

1. Substrate of chitin being solubilized by deacetylation. Then the derived substrate is broken by chitinase enzyme. Chitinase hydrolyses the glycosidic linkages of chitin [1, 2].

2 Materials

1. *Chitin agar medium*: Colloidal chitin 0.5 g; 0.07 g K_2HPO_4, 0.03 g KH_2PO_4, 0.05 g $MgSO_4$, 0.2 g NH_4NO_3, 0.1 g NaCl, 0.1 g trace elements. pH 7.8.
2. *Preparation of colloidal chitin*: 5 kg of chitin from crab shells/ any waste resources rich with chitin added to 100 ml of concentrated cold HCl with mild agitation at 4 °C for 24 h. Then, add 500 ml of ice cold ethanol (96%) to this, left 24 h with rapid stirring at 4 °C. Harvest the precipitate by centrifuge at 12,000 × *g* for 25 min at 4 °C. Wash repeatedly with sterile water till the pH reaches 6. The colloidal chitin stored at 4 °C, approximately 96 g of colloidal chitin obtained. The preparation of colloidal chitin procedure slightly varies according to the sources.

M. Senthilkumar, N. Amaresan and A. Sankaranarayanan, *Plant-Microbe Interactions: Laboratory Techniques*, Springer Protocols Handbooks, https://doi.org/10.1007/978-1-0716-1080-0_53,

3 Methods

1. Collect the soil sample from hot environmental region (temperature ranges from 45–55 °C.
2. For primary screening of thermostable chitinase producing bacteria, sample inoculated with chitin agar medium.
3. Keep the plates for incubation at 60 °C for 3 days.
4. Observe the halo zone around the colonies, confirm the presence of chitinase enzyme producing bacteria.
5. Transfer that potential isolate (s), fresh chitin containing broth medium, incubate at 60 °C.
6. Preserve the strain in glycerol stock, for further analysis.

4 Observation

1. Observe the halo zone around the bacteria after the incubation period.

References

1. Roberts WK, Seitrennikoff CP (1988) Plant and bacteria chitinases differ in antifungal activity. J Gen Microbiol 134:169–176
2. Senol M, Nadaroglu H, Dikbas N, Kotan R (2014) Purification of chitinase enzymes from *Bacillus subtilis* bacteria TV-125, investigation of kinetic properties and antifungal activity against *Fusarium culmorum*. Ann Clin Microbiol Antimicrob 13(35):1–7

Check for updates

Chapter 54

Isolation of Bacterial Hydrolytic Enzyme: Chitinase

Abstract

Microorganisms especially bacteria produced different hydrolytic enzymes and involved in defense mechanisms of plants. Microbial chitinases involved in weakening and degrading many pathogens and exhibited antibacterial, antifungal, insecticidal, or nematicidal activities.

Keywords Chitinase, Bacteria, Chitin agar, Chitinolytic

1 Principle

1. As described earlier.

2 Materials

1. Agar medium amended with chitin: Na_2HPO4 6 g; NH_4Cl 1 g; NaCl 0.5 g; Yeast extract 0.05 g; agar 15 g, and colloidal chitin.

3 Methods [1, 2]

1. Collect the soil samples from rhizospheric plants.
2. For isolation of chitinase producing bacteria, agar medium is amended with colloidal chitin.
3. Inoculate the soil sample on the surface of agar medium amended with colloidal chitin in petri plates.
4. Keep the plates for incubation for 3–4 days at 37 °C.
5. Isolates producing large hydrolytic zones are taken into consideration.
6. That isolates transferred into broth medium, after the period of incubation centrifuge at 168 g for 15 min at 4 °C.
7. Collect the crude and use for further analysis.

M. Senthilkumar, N. Amaresan and A. Sankaranarayanan, *Plant-Microbe Interactions: Laboratory Techniques*, Springer Protocols Handbooks, https://doi.org/10.1007/978-1-0716-1080-0_54,

4 Observation

1. Observe the halo zone around the bacteria after the incubation period.

References

1. Hsu SC, Lockwood JL (1975) Powdered chitin as a selective medium for enumeration of actinomycetes in water and soil. Appl Microbiol 29:422–426
2. Wiwat C, Siwayaprahm P, Bhumiratana A (1999) Purification and characterization of chitinase from *Bacillus circulans*. Curr Microbiol 39:134–140

Chapter 55

Isolation of Fungal Hydrolytic Enzyme: Chitinase

Abstract

Fungi have played an important role in biocontrol of insect pests. The chitinase enzyme is produced by fungi to degrade the substrate as a defense mechanism and quench nutrients for their survival under harsh environmental conditions. Different chitinase isomers are produced by fungi which exhibited different catalytic actions to breakdown the chitin.

Keywords Chitinase, Hydrolytic enzyme, Catalysis, Colloidal chitin agar

1 Principle

1. As described earlier (Chapter 53).

2 Materials

1. Potato dextrose agar medium: Dextrose 20 g/l; agar 18 g/l; potato starch 4 g/l; pH 5.6; 1 gram streptomycin.
2. Colloidal chitin agar medium: KH_2PO_4 2 g/l; $MgSO_4$ 0.3 g/l; $(NH_4)_2SO4$ 1.4 g/l; $CaCl_2$0.5 g/l; Peptone 0.5 g/l, Urea 0.3 g/l; $FeSo_4$ 0.005 g/l, $MnSO_4$ 0.0016 g/l, $ZnSO_4$ 0.0013 g/l, $CoCl_2$ 0.002 g, agar 15 g/l, streptomycin 1 g/l; pH 6.

3 Methods

1. Collect the soil samples from rhizospheric plants and transferred to lab in sterile polythene bags.
2. Perform serial dilution and inoculate the soil sample into potato dextrose agar medium with streptomycin antibiotic to inhibit the growth of bacteria. Incubation period 3–5 days at room temperature. Observe the fungal growth.

M. Senthilkumar, N. Amaresan and A. Sankaranarayanan, *Plant-Microbe Interactions: Laboratory Techniques*, Springer Protocols Handbooks, https://doi.org/10.1007/978-1-0716-1080-0_55, © Springer Science+Business Media, LLC, part of Springer Nature 2021

3. For isolation of chitinase producing bacteria, agar medium amended with colloidal chitin.
4. Inoculate the fungal isolate on the surface of agar medium amended with colloidal chitin in petri plates.
5. Keep the plates for incubation for 3–5 days at room temperature.
6. Isolates producing large hydrolytic zones around the colonies taken into consideration for further process.

4 Observation

1. Observe the halo zone around the fungi after the incubation period in the colloidal chitin agar medium.

Check for updates

Chapter 56

Characterization of Chitinase Enzyme from *Streptomyces*

Abstract

Plants have no immune system and contradict with vertebrates. Hence the plant communities are easily attacked by many pathogens. Hence the plant defense mechanisms by various plant growth support microbes played a significant role in the defense mechanisms against pathogens. In general, the activities of enzymes depend on the pH, temperature, and other parameters. The characterization of partially purified/purified chitinase enzyme checked with various temperature, pH, metal ions, and substrates narrated in this protocol.

Keywords Chitin, Characterization, *Streptomyces*, Optimum

The characterization of partially purified/purified chitinase enzyme checked with various temperature, pH, metal ions, and substrates.

1 Materials

1. Preparation of colloidal chitin: 5 g of chitin powder acidified with 60 ml of concentrated hydrochloric acid and kept for 2 h. Then add ice-cold distilled water until obtaining the white precipitate and it is filtered and the filtrate is washed again repeatedly with ice-cold distilled water until it reached pH 5. Then the suspension dried at hot air over for 60 °C. The dried chitin precipitate weighed and suspended in distilled water to make 4–5% colloidal chitin solution and autoclaved at 15 lbs. pressure for 15 min.

2 Methods [1, 2]

1. The characterization of partially purified/purified chitinase enzyme to be checked in temperature, pH, metal ions, and substrate.
2. Incubate 400 μl of partially purified/purified chitinase enzyme with 0.5% w/v colloidal chitin and N-acetyl glucosamine (NAG 50 μ mol) in different incubation temperature at

M. Senthilkumar, N. Amaresan and A. Sankaranarayanan, *Plant-Microbe Interactions: Laboratory Techniques*, Springer Protocols Handbooks, https://doi.org/10.1007/978-1-0716-1080-0_56,

20, 30, 40, 50, 60, 70, 80 ° C separately. Assay the mixtures for chitinase activity.

3. 400 μl of partially purified/purified chitinase enzyme incubate with 0.5% w/v colloidal chitin. Add 400 μl citrate buffer for pH 5 and 6, then 400 μl of potassium phosphate buffer for pH 7 and Tris Cl for pH 8 for 2 h separately. Assay the mixtures for chitinase activity.
4. Incubate different metal ions (10 mM) with 400 μl of partially/purified chitinase enzyme. Assay the mixtures for chitinase activity.
5. Substrates like carboxymethyl cellulose, colloidal chitinin, starch, xylan tested to determine the substrate binding ability with chitinase.
6. For this add 5 ml of partially purified/purified chitinase enzyme to 5 ml of 1% (w/v) of carboxymethyl cellulose, colloidal chitinin, starch, xylan separately and incubate at 4 ° C with mild agitation for 24 h.
7. After centrifugation at 11,000 × *g* for 20 min, discard the supernatant, wash the pellet with sterile distilled water to remove any unbound contaminated proteins.
8. Then, elute the chitinase enzyme by incubation at 30 °C with gentle agitation. Centrifuge the sample at 11,000 × g for 20 min, discard the pellet. 200 μl of supernatant taken out at 0 h, 30 min, 1 h, 2 h, 6 h, 12 h, and 24 h and assay the chitinase enzyme by the release of NAG.

References

1. Ray L, Panda AN, Mishra SR, Pattanaik AK, Adhya TK, Suar M, Raina V (2018) Purification and characterization of extra cellular thermo-alkali stable, metal tolerant chitinase from *Streptomyces chilikensis* RC 1830 isolated from a brackish water lake sediment. Biotechnol Rep 20:1–7
2. Zarei M, Aminzadeh S, Zolgharnein H, Safahieh A, Daliri M, Noghabi AK, Ghoroghi A, Motallebi A (2011) Characterization of a chitinase with antifungal activity from a native *Serratia marscens* B4A. Braz J Microbiol 42:1017–1029

Check for updates

Chapter 57

Chitinase Assay by Spectrophotometric Method

Abstract

The chitin degrading enzyme chitinase is the most abundant polymer next to cellulose. Chitin is present in the cell wall of fungi. Chitinase hydrolyzes β-1,4 glycosidic bonds between the N-acteyl glucosamine residue of chitin. Certain setbacks with the existing chitinase assay method are either time bounding or less sensitivity in the assay methods. Spectrophotometric method is an important method in the assay of chitinase.

Keywords Chitinase assay, Spectrophotometer, Fungi, Hydrolysis

1 Materials

1. Colloidal chitin.
2. DNS reagent.

2 Methods [1]

1. For the chitinase assay, take the colloidal chitin as a substrate and add with partially purified/purified chitinase enzyme.
2. Add 0.9 ml of 1% (w/v) colloidal chitin with 0.1 ml of crude enzyme, incubate at 55 °C for 1 h.
3. By the addition of 3 ml of DNS, stop the reaction then heat at 100 °C for 5 min.
4. Centrifuge it, estimate the reducing sugar by DNS method.
5. Dilute the sample, measure the absorption at 530 nm using UV spectrophotometer along with the substrate and enzyme blanks.

Reference

1. Vahed M, Motalebi E, Rigi G, Noghabi KA, Soudi MR, Sedeghi M, Ahmadian G (2013) Improving the chitinolytic activity of *Bacillus pumilus* SG 2 by random mutagenesis. J Microbiol Biotechnol 23(11):1519–1528

M. Senthilkumar, N. Amaresan and A. Sankaranarayanan, *Plant-Microbe Interactions: Laboratory Techniques*, Springer Protocols Handbooks, https://doi.org/10.1007/978-1-0716-1080-0_57,

Check for updates

Chapter 58

Extraction of Fungal Chitinase Enzyme

Abstract

Extraction is an important part in any enzyme. The extraction procedures revealed various optimization conditions. The extraction procedure varies between one enzyme and the other. In general, the extraction conditions are controlled by various parameters like pH, substrate, time and temperature, and the concentration of enzymes. In this protocol, the extracted fungal chitinase is measured by absorbance in spectrophotometer. The readings are compared with the standard.

Keywords Chitinase, Extraction, Fungi, Enzyme

1 Materials

1. Potato Dextrose agar medium: Dextrose 20 g/l; agar 18 g/l; potato starch 4 g/l; pH 5.6; 1 g streptomycin.
2. 10x phosphate buffered saline (PBS, 1.35 M NaCl, 47 mM KCl, 100 mM Na2HPO4, 20 mM NaH_2PO_4, pH 7.4)
3. Mcllvaine's buffer: 0.2 M Na_2HPO_4.
4. M citric acid (pH = 6.0).
5. Ehrlich's solution.
6. 10 g *p*-dimethyl amino benzaldehyde in 12.5 ml concentrated HCl (37%) and 87.5 ml glacial acetic acid
7. 0.27 M sodium borate.

2 Methods [1]

1. Isolate the potential chitinase enzyme producing fungi after screening process from potato dextrose agar plates.
2. Cut the agar culture into squares and the squares are cultured into liquid medium for another 2–3 days.
3. Filter the mycelium and collect the mycelium, then lyophilize the mycelium. Mix 5 mg of mycelium with 1 ml of 6% KOH in

M. Senthilkumar, N. Amaresan and A. Sankaranarayanan, *Plant-Microbe Interactions: Laboratory Techniques*, Springer Protocols Handbooks, https://doi.org/10.1007/978-1-0716-1080-0_58,

each 2 ml microtube using a vortex mixer, then incubate in a water bath at 80 °C for 90 min.

4. Centrifuge the sample at 24,000 g for 10 min and discard the suspension.
5. Wash each pellet with 1 ml 1 × phosphate buffered saline (PBS) for three times then centrifuge at 12,500 rpm for 5 min to discard the suspension.
6. Resuspend the pellet with 0.5 ml Mcllvaine's buffer. Add 100 μl chitinase into each sample and incubate at 37 °C for 16 h in the dark at 220 rpm in an incubator shaker.
7. Chitinase treated samples are mixed with equal volume of 0.27 M sodium borate (pH = 9.0) and incubate at 100 °C for 10 min.
8. 200 μl of each sample is added to 1 ml Ehrlich's solution after cooling down to room temperature, then incubate at 37 °C for 30 min in an incubator.
9. Transfer 100 μl of each sample into a microtiter plate and measure the absorbance at 585 nm using a microplate reader. Standard curves are prepared using stock of N-acetyl glucosamine with different concentration.

Reference

1. Bulik DA, Olczak M, Lucero HA, Osmond BC, Robbins PW, Specht CA (2003) Chitin synthesis in *Saccharomyces cerevisiae* in response to supplementation of growth medium with glucosamine and cell wall stress. Eukaryot Cell 2 (5):886–900

Check for updates

Chapter 59

Partial Purification of Bacterial Chitinase Enzyme by Ion Exchange Chromatography

Abstract

Purification is an important step in enzyme production which helped in removing the impurities from the enzyme. The maximum activity is exhibited after purification step. Ion exchange chromatographic technique is widely used to chromatographic technique for the separation and purification of proteins, biomolecules, and nucleic acids. This technique is functioning based on the isoelectric point of an enzyme.

Keywords Chitinase, Isoelectric point, Ion exchange chromatogram

1 Materials

- M phosphate buffer solution (pH 7.0): Take 800 ml of distilled water, add 15.487 g of $Na_2HPO_4 \cdot 7H_2O$ to the solution. Then add 5.827 g $NaH_2PO_4 \cdot H_2O$ to the solution. Adjust the solution to pH 7 using HCl or NaOH. Bring the final volume to 1 L with distilled water.
- M Tris–HCl buffer solution (pH 7.0): Take 121.14 g of Tris in 800 ml of distilled water, then adjust the pH 7 using HCl. Bring the final volume to 1 L with distilled water.
- M sodium citrate solution (pH 6.0): Take 24.096 g of sodium citrate dehydrate into 800 ml of distilled water then add 3.471 g citric acid into the solution. Adjust the pH using HCl or NaOH solution. Bring the final volume to 1 L with distilled water.
- M NaCl solution: Take 58.44 g of NaCl in 1 L of water.

2 Methods [1]

1. Homogenate the chitinase enzyme producing bacteria (screened from isolation procedure) and the precipitate the homogenate by ammonium sulfate precipitation at 0–100%.
2. Dissolve the precipitation in 0.1 M phosphate buffer (pH 7.0), dialyzed it against the same buffer.

M. Senthilkumar, N. Amaresan and A. Sankaranarayanan, *Plant-Microbe Interactions: Laboratory Techniques*, Springer Protocols Handbooks, https://doi.org/10.1007/978-1-0716-1080-0_59, © Springer Science+Business Media, LLC, part of Springer Nature 2021

3. Allow the chitinase enzyme to stand at −25 °C in 0.1 M Tris–HCl buffer (pH 7.0).
4. Apply the dialyzed suspension into previously equilibrated DEAE Sephadex ion exchange column (2.5 × 30 cm) with 20 mM sodium citrate (pH 6.0).
5. Wash the column with same buffer.
6. Elute the bound proteins by applying a gradient to the column from 0 to 1 M NaCl and collect the fractions with a 3 ml/min. Flow rate.
7. Determine the activity of partially purified chitin with the colloidal chitin substrate.

Reference

1. Senol M, Nadaroglu H, Dikbas N, Kotan R (2014) Purification of chitinase enzymes from *Bacillus subtilis* bacteria TV-125, investigation of kinetic properties and antifungal activity against *Fusarium culmorum*. Ann Clin Microbiol Antimicrob 13:351–357

Check for updates

Chapter 60

Isolation of Actinomycetes Producing Hydrolytic Enzyme: β-1,3 Glucanase

Abstract

β-1,3 glucanase is a hydrolytic enzyme that cleaves the glycosidic bonds adjacent to glucose residues in β-1,3 glucan. β-1,3 glucan, is an important component of endosperm of plant cell wall. β-1,3 and β-1,4 glucanases are widely present in higher plants, bacteria, and fungi. The microbial β-1,3 glucanase enzymes are having wide attraction especially actinomycetes; it is easy to cultivate and maintain.

Keywords β-1,3 glucanase, Actinomycetes, Hydrolytic enzyme

1 Principle

β-1,3 glucanase enzyme hydrolyses β-1,3 glucosidic bonds of β-1,3 glucan. Hydrolases enzymes are involved in this activity [1].

2 Materials

Yeast casaminoacids extracts and dextrose agar (YCED).

3 Methods

1. Collect the soil sample and inoculate on YCED agar plates.
2. Keep the plate for incubation at 28 °C for a week period.
3. After the period of incubation, observe the plates for halo zone formation.
4. The isolates producing wider zones indicated the strong activity of β-1,3 glucanase.
5. Preserved the potential isolates for further analysis.

M. Senthilkumar, N. Amaresan and A. Sankaranarayanan, *Plant-Microbe Interactions: Laboratory Techniques*, Springer Protocols Handbooks, https://doi.org/10.1007/978-1-0716-1080-0_60,

4 Observations

Observe the halo zone around the bacteria after the incubation period.

Reference

1. Wu Q, Dou X, Wang Q, Guan Z, Cai Y, Liao X (2018) Isolation of β 1,3 glucanase producing microorganism from *Poria cocos* cultivation soil via molecular biology. Molecules 23:1–19

Check for updates

Chapter 61

Isolation of Bacteria Producing Hydrolytic Enzyme: β-1,3 Glucanase

Abstract

β-1,3 glucanase is a hydrolytic enzyme that cleaves the glycosidic bonds adjacent to glucose residues in β-1,3 glucan. The bacterial β-1,3 glucanase enzyme has a lot of advantages. It is highly temperature resistant and easy for genetic modification than plant-based β-1,3 glucanase. Regarding fungi, β-1,3 glucanase are mostly acidic and working well in low pH.

Keywords β-1,3 glucanase, hydrolytic enzyme, bacteria

1 Principle

- As described earlier (Chapter 60).

2 Materials

- Selective medium: 0.3 g barley glucan; 0.1 g Congo red; 1 g NaCl; agar 2 g.

3 Methods [1]

1. Collect the moldy cereals containing booklice using sterile bag and isolate the booklice using floatation technique.
2. Ground the booklice specimens and collect the specimens using 1 ml micro dismembrator.
3. Dilute the specimens and spread on the selective medium.
4. Keep the plate for incubation at 37 °C for 24 h.
5. After the period of incubation, observe the plates for hydrolysis zone formation.

M. Senthilkumar, N. Amaresan and A. Sankaranarayanan, *Plant-Microbe Interactions: Laboratory Techniques*, Springer Protocols Handbooks, https://doi.org/10.1007/978-1-0716-1080-0_61,

6. The isolates producing wider zones indicated the strong activity of β-1,3 glucanase.
7. Preserve the potential isolates for further analysis.

4 Observation

- Observe the halo zone around the bacteria after the incubation period.

Reference

1. Niu Q, Zhang G, Zhang L, Ma Y, Shi Q, Fu W (2016) Purification and characterization of a thermophilic 1,3-1,4-β glucanase from *Bacillus methylotrophicus* S2 isolated from book lice. J Biosci Bioeng 121(5):503–508

Check for updates

Chapter 62

Purification of *Trichoderma* Producing Hydrolytic Enzyme: β-1,3 Glucanase

Abstract

Hydrolytic enzyme β-1,3 glucanase cleaves the glycosidic bonds adjacent to glucose residues in β-1,3 glucan. This enzyme is inhibiting the growth of the fungi. In the present protocol, the purification of *Trichoderma* producing β 1,3 glucanase is narrated.

Keywords β-1,3 glucanase, *Trichoderma*, β-1,3 glucan

1 Principle

- As described earlier (Chapter 60).

2 Materials

- *Trichoderma* liquid enzyme production (TLE) medium: (9–479): 0.1 g (w/v) bactopeptone; 0.03 g urea; 0.2 g KH_2PO_4; 0.14 g $(NH_4)_2SO_4$ 0.03 g; $MgSO_4$; 2 g trace elements solution containing 0.025 g $FeSO_4$; 0.0085 g $MnSO_4$; 0.007 g $ZnSO_4$; 0.01 g $CaCl_2$; supplemented with 0.5 g *F. solani* cell wall or with 2 g glucose; pH of the medium adjusted to 5.
- Malt Yeast Glucose (MYG) medium: 0.5 g Malt yeast extract; 0.25 g yeast extract; 1 g glucose, and 2 g agar.
- Sephacryl S-100 column.

3 Methods [1, 2]

1. Cultivate the *T. harzianum* or *T. asperellum* (obtained from the culture collection center) in the MYG medium, harvest the spores (1×10^7 spores/ml) and collected it in sterile water.

M. Senthilkumar, N. Amaresan and A. Sankaranarayanan, *Plant-Microbe Interactions: Laboratory Techniques*, Springer Protocols Handbooks, https://doi.org/10.1007/978-1-0716-1080-0_62,

2. Centrifuged at 1000 rpm, wash twice and inoculate in 200 ml TLE medium placed in 1 L Erlenmeyer flask.
3. Grow the culture in orbital shaker at 180 rpm at 28 °C for 48 h.
4. Harvest the mycelium by filtration through filter paper, the culture filtrate dialyzed over night against distilled water, freeze dried.
5. It may be used as a source of β-1,3 glucanase enzyme.
6. Load the enzyme sample in Sephacryl S-100 column, previously equilibrated with 20 mmol Tris–HCl buffer at pH 8, elute with same buffer with a flow rate of 40 ml/h.
7. Fractions were collected and monitored for β-1,3 glucanase activity.

References

1. Bara MTF, Lima AL, Ulhoa CJ (2003) Purification and characterization of an exo- β 1,3-glucanase produced by *Trichoderma asperellum*. FEMS Microbiol Lett 219:81–85
2. Ramada MHS, Steindorff AS, Bloch C Jr, Ulhoa CJ (2016) Secretome analysis of the mycoparasitic fungus *Trichoderma harzianum* ALL 42 cultivated in different media supplemented with *Fusarium solani* cell wall or glucose. Proteomics 16:477–490

Check for updates

Chapter 63

Estimation of Lipoxygenase

Abstract

Lipoxygenase catalyzes the dioxygenation of polyunsaturated fatty acid in lipids. Further, it is involved in various signaling mechanisms and regulates the function of the cells. This protocol narrates the estimation of lipoxygenase enzyme.

Keywords Lipoxygenase, Lipids, Catalytic enzyme

1 Principle

Lipoxygenase (EC 1.13.11.12) is also called as lipoxidase, which is a dioxygenase that catalyzes the hydroperoxidation by molecular oxygen of linoleic acid and other polyunsaturated lipids. The formation of the conjugated diene from linoleate is measured at 234 nm.

2 Materials

- Acetone.
- Diethyl ether.
- 1 M Tris–HCI buffer containing 0.1% Triton X–100 at pH 7.3.
- Absolute ethanol.
- Tween 20.
- 0.05 M Na_2HPO_4.
- 1 N NaOH.

Substrate

- Stock A: 1% linoleic acid (w/v) in absolute ethanol.
- Stock B: To 7.1 ml of stock A, add 0.25 ml of Tween 20. Evaporate the ethanol under vacuum in a rotary evaporator. Dissolve the residue in 100 ml of Na_2HPO_4 and adjust the pH to 9.0 using 1 N NaOH. This stock solution contains

M. Senthilkumar, N. Amaresan and A. Sankaranarayanan, *Plant-Microbe Interactions: Laboratory Techniques*, Springer Protocols Handbooks, https://doi.org/10.1007/978-1-0716-1080-0_63,

linoleic acid and Tween at levels of 2.5 c 10^{-3} M and 0.25%, respectively.

3 Methods

1. Cool the material in a dry ice bath (−30 °C) and grind to a fine powder.
2. Defat with acetone and diethyl ether and suspend 1.0 g of defatted material in 3 volumes of Tris–HCI buffer with mechanical stirring for 16 h.
3. Centrifuge at 48,000 × *g* for 30 min and discard the pellet.
4. Subject the supernatant to two additional centrifugation steps (30 min, 48,000 × *g*). Discard the pellet.
5. Resubject the supernatant to three repeated ultracentrifugation at 200,000 × *g* for 3 h and discard the pellet. Use the supernatant as enzyme source.

4 Enzyme Assay

1. Dilute stock solution B tenfold with 0.2 M citrate-phosphate buffer of the desired pH (the pH of the final substrate solution is that of the diluting buffer).
2. Take 2.4 ml of substrate solution and 0.1 ml of water in the control cuvette.
3. Zero the UV recording spectrophotometer with both reference and sample cuvette filled with the control solution.
4. Empty the sample cuvette and fill with 2.4 ml of substrate solution. At zero time add 0.1 ml of enzyme extract, mix rapidly and start the recorder so that the time.
5. Interval between addition of the enzyme and start of the recorder is not more than 10 s.
6. Compute the rate of increase of optical density (OD234/min) from the initial linear potion of the graph.
7. Express the enzyme activity as OD234/min/mg protein or per g weight of material.

Part V

Techniques for Studying Biological Control of Insects and Nematodes

- Blood agar.
- Water bath.
- Circular rotator.

3 Methodology

1. Prepare the culture of obligate anaerobes in logarithmic growth phase by inoculating pre-reduced thioglycolate broth; then incubate the culture at 35 °C in an anaerobic glove box.
2. Use 12-h-old culture for study after four successive transfers in the above medium.
3. To obtain large number of spores, incubate the *Bacillus* cultures in cooked meat broth for 5 days at ambient temperature.
4. Mix the culture thoroughly after incubation.
5. Use 1 mL aliquots of the cooked meat and thioglycolate culture for heat shock and ethanol treatment.
6. *Heat shock treatment*: Heat 1 mL aliquots of cooked-meat broth and thioglycolate cultures at 80 °C water bath for 15 min and then transfer it to a cool water bath at 25 °C for 30 min.
7. *Alcohol treatment*: Rotate an aliquot (0.5 mL) of a cooked-meat or thioglycolate broth culture and 0.5 mL of absolute ethanol in screw-cap tubes at 35 rpm on a circular rotator for 1 h at room temperature.
8. Perform the serial dilution to 10^{-6} of each heat-treated, ethanol-treated and untreated culture by preparing ten-fold dilutions in pre-reduced buffered gelatin.
9. Dilute the anaerobic cultures in anaerobic medium.
10. Inoculate 0.1 mL of the 10^{-2}, 10^{-4}, 10^{-6} dilutions and undiluted culture on to the anaerobe blood agar plates.
11. Spread the inoculum carefully.
12. Incubate the plates at 35 °C in an anaerobic system for quantitation of obligatory anaerobic organisms and in candle extinction jars for the facultative anaerobic and aerobic bacteria.
13. Count the colonies after 48–72 h of incubation.
14. Perform the experiment in triplicates and calculate the average.
15. Calculate the CFU/mL of the heat-shocked, ethanol treated and untreated culture.

4 Note

- For isolating the spores, determine the most effective ethanol concentration and exposure time by varying the time in the rotator (15, 30, 45, and 60 min) and the concentration of absolute ethanol (10%, 25%, 50%, and 70%) used in the procedure.

Reference

1. Koransky JR, Allen SD, Dowell VR (1978) Use of ethanol for selective isolation of spore forming microorganisms. Appl Environ Microbiol 35:762–765

Check for updates

Chapter 65

Spore Staining and Biochemical Characterization of *Bacillus*

Abstract

Physiological characters of bacteria differ from one species to other species. Therefore, biochemical tests are used to identify various types of *Bacillus* species on the basis of their difference in biochemical activity such as production of enzymes, carbohydrate metabolism, fat metabolism, or an ability to produce a particular compound. Spore staining is used to identify the type of bacterial endospore. Both spore staining and biochemical characterization are the preliminary basis of identification.

Keywords *Bacillus*, Spore, Biochemical test, Spore stain, Carbohydrate utilization

The species, *Bacillus thuringienesis* is distinguished by its bio-insecticidal activity against some orders of insects such as Lepidoptera, Diptera, and Coleoptera. *B. thuringiensis* (Bt) is ubiquitous, facultative aerobic, gram-positive, rod-shaped spore forming bacterium that forms parasporal crystals during its sporulation process which are mainly responsible for its insecticidal activity. Vegetative cell is of 1.0–1.2 μm wide and 3.0–5.0 μm in length and usually motile by peritrichous flagella. Temperature range of Bt is 30–45 °C. The spore of this bacterium is of ellipsoidal shape or cylindrical and is located centrally or paracentrally inside the mother cell. The spores are resistant to starvation, ultraviolet radiation, heat, desiccation, freezing, chemical agents, and commonly used staining technique.

1 Spore Staining

Some of the staining methods commonly used to stain the spores include Dorner's method, Scheffer and Fulton's method, Sydner's modification of Dorner's method, Fleming's method, and Bartholomew and Mittwer's method. But the widely used differential staining procedure used to stain the endospore is Scheffer and Fulton's method. In this method, the primary stain, malachite green is forced into the spore by steaming the bacterial emulsion.

M. Senthilkumar, N. Amaresan and A. Sankaranarayanan, *Plant-Microbe Interactions: Laboratory Techniques*, Springer Protocols Handbooks, https://doi.org/10.1007/978-1-0716-1080-0_65,

Malachite green is water soluble and has low affinity for cellular material, so vegetative cells will be decolorized with water. Vegetative cells are then counter-stained with safranin and observed under microscope.

2 Biochemical Characterization

Biochemical characterization is routinely used to distinguish between closely related *Bacillus* species. Some of the biochemical tests which are mostly given positive by most of the Bt strain includes utilization of carbohydrates, Voges–Proskauer test, nitrate reduction test, citrate utilization test, hydrolysis of starch and casein, liquefaction of gelatin, decomposition of tyrosine, deamination of phenylalanine, catalase reaction, Egg-yolk reaction, and resistance to lysozyme. Follow the standard protocols for all these biochemical reaction described by Gordon et al. [1] or from the procedure described elsewhere. For all biochemical tests, it is essential that results must be verified using reference strains as positive and/or negative control [2]. The methodology described by Gordon et al. [1] for the identification of *Bacillus* can now be considered valuable in identifying spore forming *Bacillus* across the many genera.

Biochemical characteristics of B. thuringiensis

S. No.	Biochemical tests	*Bacillus thuringienesis*
1	Utilization of carbohydrates	
	(a) D-Glucose	+
	(b) D-Mannitol	–
	(c) D-Mannose	–
	(d) D-Xylose	–
	(e) Sucrose	–
	(f) Glycogen	+
	(g) Salicilin	–
2	Voges–Proskauer test	+
3	Nitrate reduction test	+
4	Citrate utilization test	+
5	Hydrolysis of starch	+
6	Hydrolysis of casein	+
7	Liquefaction of gelatin	+
8	Decomposition of tyrosine	+
9	Deamination of phenylalanine	–
10	Catalase reaction	+
11	Egg-yolk reaction	+
12	Resistance to lysozyme	+

2.1 Note

- Results may vary depending upon the Bt strain.

3 PCR-Based Screening for Identification of *Bacillus* with Crystal Protein

To identify the specific *cry* and *cyt* genes by PCR, the specific primers pairs are used that specifically recognize the single *cry* gene and yield an amplicon that can be visualized by agarose gel electrophoresis. A mixture of more than two primers (multiplex PCR) in the same PCR reaction mixture can also be employed. In this situation, the PCR amplification yields as many different sizes of amplicons. These amplicons are easily identified on the basis of their size and determined by the agarose gel electrophoresis [3–5].

3.1 Materials

- dNTPs.
- Taq polymerase.
- 10× Buffer with $MgCl_2$.
- Nuclease free water.
- Universal primer set for *Bacillus thuringienesis* or other insecticidal *Bacillus* species.
- Specific primer set for crystal protein genes.
- Agarose gel.
- 1× TAE buffer
- Gel electrophoretic unit.
- PCR thermal cycler.

3.2 Methodology

1. Extract the genomic DNA from crystal protein producing *Bacillus* species with standard protocol.
2. Analyze the total extracted DNA on 1% agarose gel using ethidium bromide.
3. Use genomic DNA isolated from *Bacillus* strain as a DNA template and following pairs of primers for PCR amplification to screen out different *cry* type genes.

Sr. No.	Name	Sequence (5′ → 3′)	Amplification	Product size
1.	Un1F	CATGATTCA TGCGGCAGATAAAC	Partial *cry*1 gene	277 bp
	Un1R	TTGTGACACTTCTGC TTCCCATT		
2.	Un2F	GTTATTCTTAATGCAGA TGAATGGG	Partial *cry*2 gene	689–701 bp
	Un2R	CGGATAAAATAATC TGGGAAATAGT		

(continued)

Sr. No.	Name	Sequence (5' → 3')	Amplification	Product size
3.	Un3F Un3R	CGTTATCGCAGAGAGATGACATTAAC CATCTGTTGTTTCTGGAGGCAAT	Partial *cry*3 gene	589–604 bp
4.	Un4F Un4R	GCATATGATGTAGCGAAACAAGCC GCGTGACATACCCATTTCCAGGTCC	Partial *cry*4 gene	439 bp
5.	Un5F Un5R	TTACGTAAATTGGTCAATCAAGCAAA AAGACCAAATTCAATACCAGGGTT	Partial *cry*5, 12, 14, 21 genes	474–489 bp
6.	Un7–8F Un7–8R	AAGCAGTGAATGCCTTGTTTAC CTTCTAAACCTTGACTACTT	Partial *cry*7–8 gene	420 bp
7.	Un9F Un9R	CGGTGTTACTATTAGCGAGGGCGG GTTTGAGCCGCTTCACAGCAATCC	Partial *cry*9 genes	351–359 bp
8.	Un11F Un11R	TTCCAACCCAACTTTCAAGC AGCTATGGCCTAAGGGGAAA	Partial *cry*11 genes	305 bp
9.	VipF VipR	CCTCTATGTTGAGTGATGTA CTATACTCCGCTTCACTTGA	Partial *vip*3 genes	1000 bp
10.	Cyt1F Cyt1R	AACCCCTCAATCAACAGCAAGG GGTACACAATACATAACGCCACC	Partial *cyt*1 genes	522–525 bp
11.	Cyt2F Cyt2R	AATACATTTCAAGGAGCTA TTTCATTTTAACTTCATATC	Partial *cyt*2 genes	469 bp

4. Prepare PCR reaction mixture in 50 μl by adding 200 mM each deoxynucleotide triphosphate, 2.5 U of Taq DNA polymerase, 0.5 mM each forward and reverse primer, DNA template, and buffer.

5. Setup PCR conditions in thermal cycler for amplification: 5 min denaturation at 94 °C followed by 35 cycles of amplification with 1 min denaturation at 94 °C, 45 s of annealing at 45–55 °C (depending upon cry gene primer) and 2 min

extension at 72 °C followed by final extension step at 72 °C for 10 min.

6. Analyze the PCR products by 1% agarose gel electrophoresis technique.

References

1. Gordon R, Haynes WC, Pang CHN (1973) The genus *Bacillus*. In: Agriculture handbook no. 42. United States, Department of Agriculture, Washington, DC
2. Rabinovitch L, Vivoni AM, Machado V, Knaak N, Berlitz DL, Polanczyk RA, Fiuza LM (2017) *Bacillus thuringiensis* characterization: morphology, physiology, biochemistry, pathotype, cellular, and molecular aspects. In: *Bacillus thuringiensis and Lysinibacillus sphaericus*. Springer, Cham, pp 1–18
3. Juárez-Pérez VM, Ferrandis M, Frutos R (1997) PCR-based approach for detection of novel *Bacillus thuringiensis* cry genes. Appl Environ Microbiol 63:2997–3002
4. Jain D, Kachhwaha S, Jain R, Kothari SL (2012) PCR based detection of cry genes in indigenous strains of *Bacillus thuringiensis* isolated from the soils of Rajasthan. Indian J Biotechnol 11:491–494
5. Jain D, Sunda SD, Sanadhya S, Nath DJ, Khandelwal SK (2017) Molecular characterization and PCR-based screening of cry genes from *Bacillus thuringiensis* strains. 3 Biotech 7:4

Check for updates

Chapter 66

Insect Bioassay for Determining LD_{50} of Crystal Protein

Abstract

Amino acid sequences and insecticidal spectrum are the major criteria used to classify crystal protein gene of *Bacillus thuringiensis*. A biochemical mechanism of toxicity study reveals that *B. thuringiensis* crystal proteins form the pores inside the membranes of susceptible cells. In transgenic plants and in plants associated microbes, crystal protein of *B. thuringiensis* express themselves and their expression is proved to be useful strategy in protecting the agricultural crops from the insects attack.

Keywords *Bacillus thuringiensis*, Crystal protein, Insect attack, Endotoxins, Cytolitic toxins

1 Principle

Control of pest population in agriculture usually depends on chemical synthetic insecticides. Chemical pesticides are highly expensive and extremely toxic to non-target organisms and human and animal health. Therefore, an alternative to chemical pesticide, *Bacillus thuringiensis*, could be used as a biopesticide due to its ability to produce insecticidal crystal proteins (δ-endotoxins) and cytolitic (Cyt) toxins. Cry proteins has biocidal activity with narrow spectrum as it is based on specific receptor recognition and membrane insertion of the active Cry proteins brings the paralysis of the transepithelial transport [1].

2 Materials

- Bacterial strains.
- Plasmid.
- Agarose gel.
- Kanamycin.
- Isopropyl-β-D-1-thiogalactoside (IPTG).
- Luria–Bertini medium.
- Phenyl methane sulfonyl fluoride (PMSF).
- Sodium chloride (NaCl).

M. Senthilkumar, N. Amaresan and A. Sankaranarayanan, *Plant-Microbe Interactions: Laboratory Techniques*, Springer Protocols Handbooks, https://doi.org/10.1007/978-1-0716-1080-0_66,

- Triton-X.
- Coomassie brilliant blue dye.
- Artificial diet for larvae raring [kabuli gram flour (60 g), casein (22.5 g), ascorbic acid (9 g), yeast tablets (9 g), methyl parahydroxy benzoate (MPH) (2 g), sorbic acid (1 g), sucrose (19 g), hosta-cycline (1.5 g), multivitaplex capsule (5 nos.), formaldehyde 10% (v/v) (2 ml), sugarcane shoot powder (75 g), and agar (9 g)].
- Gel electrophoretic unit.
- Thermocycler.
- Centrifuge.
- Sonicator.

3 Methodology

3.1 Bacterial Strains and Plasmid

- Isolate the cry1F gene fragment from the vector pBi-nAR and clone it in pET28a (+) expression vector for transforming in to *Escherichia coli* (DH5α).
- Grow the transformed *E. coli* in LB broth medium containing kanamycin (50 mg/L) and incubate it at 37 °C for overnight on continuous shaking at 180 rpm.

3.2 Plasmid DNA Isolation, Manipulation, and Confirmation of cry1F

- Isolate the plasmid DNA from *E. coli* (pBinARCry1F and pET28a (+)) according to alkaline lysis method and purify it by using Qiagen-tip 100 column.
- Digest the recombinant plasmid pBinAR-cry1F and pET28a (+) vectors with the Xba I and Sal I.
- Perform agarose gel electrophoresis to purify DNA fragment of cry1F gene and pET28a (+) plasmid and ligate them using T4 DNA ligase at 16° C by incubating for 5 h to achieve recombinant pET28a (+)-cry1F (7 kb) expression construct.
- Transform the ligated product in *E. coli* and colonies which are obtained under the selection of kanamycin (50 mg/L) will be subjected to colony PCR using cry1F gene specific primers (FP: 5'-ATC CAG AAT CAA TGC GTC CC-3' and RP: 5'-GAA AGA GCT CAG AAG GCG TAG-3') to amplify 1.8 kb fragment.
- The PCR mixture (20 μl) consists of pricked colony, 1× Taq buffer, 250 μM dNTPs, 1.5 mM $MgCl_2$, 0.25 μM each forward and reverse primers, and 1 U Taq DNA polymerase.
- The thermal cycler conditions would be initial denaturation at 95 °C for 4 min, followed by 35 cycle at 94 °C for 45 s of

denaturation, 59 °C for 30 s of annealing, 72 °C for 90 s of extension and final extension at 72 °C for 10 min.

- Analyze the amplified PCR products by using electrophoresis on 0.8% (w/v) agarose gel. Confirm the presence of the recombinant pET28a (+)-cry1F with restriction analysis.

3.3 Recombinant Protein (cry1F) Expression in E. coli (DH5α)

- Grow the *E. coli* strain DH5α clone harboring pET28a (+)-cry1F in 5 ml LB medium containing kanamycin (50 mg/L) at 37 °C until OD_{600} reaches to 0.6.
- Transfer the same culture into 50 ml fresh LB medium and induce it by adding isopropyl-β-D-1-thiogalactoside (IPTG) (1 mM) at 37 °C for 12 h.
- Use the *E. coli* clone harboring pET28a (+) vector as a control.
- Harvest the cells by centrifugation at 6289 × *g* for 5 min at 4 °C and wash the pellets obtained with distilled water and resuspend in 4 ml of extraction buffer (pH 8.0) containing Tris–HCl (30 mM), EDTA (1 mM) and phenyl methane sulfonyl fluoride (PMSF) (1 mM) and lyse it on ice by ultrasonication at 20 kHz with pulse of 10 s for six times with an interval of 5 s.
- Wash the pellet twice with NaCl (0.5 M) plus Triton-X (2%) followed by NaCl (0.5 M) and sterile distilled water.
- Isolate the unsolubilized protein i.e., supernatant by centrifugation at 15,600 × *g* at 4 °C for 10 min.
- Separate the expressed protein extract from *E. coli* lysate by SDS/Polyacrylamide gel (10%) electrophoresis.
- Visualize the separated polypeptides by staining with Coomassie brilliant blue and record the gel image using gel documentation unit.
- Estimate the total protein isolated from cell lysate by Bradford method, using the BSA as a standard protein.

3.4 Insect Bioassay with cry1F

- Collect the first instars larvae of insect from any of the crop plant.
- Prepare the artificial diet for larvae raring consisting of kabuli gram flour (60 g), casein (22.5 g), ascorbic acid (9 g), yeast tablets (9 g), methyl parahydroxy benzoate (MPH) (2 g), sorbic acid (1 g), sucrose (19 g), hosta-cycline (1.5 g), multivitaplex capsule (5 nos.), formaldehyde 10% (v/v) (2 ml), sugarcane shoot powder (75 g) and agar (9 g).
- Supplement the molted nutrient agar (400 ml water) with the above mixture.
- Dispense the artificial diet in glass Petri plates with varying concentration (3.14, 6.28, 9.43, 12.57 and 15.72 μg toxin/

cm^2) of partially purified cry1F proteins smeared on the surface of the diet.

- Extracted protein from *E. coli* harboring pET28a (+) vector would serve as a negative control and artificial diet without protein would serve as a control.
- Use fifteen larvae for each concentration and replicate each set thrice and record the mortality after every 3 days till 15 days after initiation.
- Determine the cumulative mortality by measuring number of dead larva per treatment as:

  Cumulative mortality (%) = (Total number of larvae survived)/(Total number of larvae released) × 100.

Reference

1. Thorat AS, Nakade HD, Sonone NA, Babu KH (2017) In vitro bioassay to determine the toxicity of cry1F protein against sugarcane early shoot borer (*Chilo infuscatellus* Snell). Int J Sci Eng Res 8:648–652

Chapter 67

Insect Bioassay for Determining LD_{50} of Spore–Crystal Protein Mixture

Abstract

Bioassay is an important method used to determine the toxicity of the spore–crystal protein mixture i.e., the lethal concentration (LC_{50}). In case of insect larvae which are susceptible, the time of death gets reduced when the spores present were used in combination with one of the purified crystal protein. In the absence of spores i.e., when insect larvae is treated with one of the crystal toxin, the intoxication speed and the eventual death will be much slower than the preparation containing mixture of both spores and crystal protein together.

Keywords Spores, Crystal protein, LC_{50}-lethal dose, Insect larvae, Spore–crystal mixture

1 Principle

The final formulation of each Bt product should be bioassayed against a specific test insect with accepted international standards. To carry out bioassay at laboratory scale different concentrations of the formulated product are compared and larvicidal activity are recorded in terms of lethal doses (LD_{50}) or lethal concentrations (LC_{50}) depending upon the method used. The potency of the formulation is defined as ITU/mg [1, 2].

2 Materials

For Preparing Spore–Crystal Mixture

- Nutrient broth.
- 0.5 M NaCl.
- 6% lactose.
- Centrifuge.
- Vacuum filtration unit.

M. Senthilkumar, N. Amaresan and A. Sankaranarayanan, *Plant-Microbe Interactions: Laboratory Techniques*, Springer Protocols Handbooks, https://doi.org/10.1007/978-1-0716-1080-0_67,

For Bioassay

- 0.1% Triton X-100.
- 0.2% Triton X-100.
- Cabbage leaf.
- Six-day-old insect larvae.
- Petri dishes.

3 Methodology

3.1 Preparation of Spore–Crystal Protein Mixture

1. Inoculate *Bacillus* strain in nutrient broth medium and incubate it on rotary shaker at 150 rpm for 96 h at room temperature.
2. Centrifuge the autolysed culture after incubation at 2795 × *g* for 10 min.
3. Discard the supernatant and resuspend the pellet in 0.5 M NaCl for 15 min.
4. Centrifuge again and wash the pellet with sterile distilled water twice.
5. Resuspend the resultant pellet after washing in 6% lactose for 30 min under constant stirring.
6. Filter it under suction, dried it overnight under partial vacuum.
7. The aliquots of the spore–crystal mixture can be stored at 4 °C for further use.

3.2 Leaf-Dip Bioassay Method for Determining LD_{50}

1. Wash the cabbage leaf thoroughly with 0.1% Triton X-100 prepared in sterile distilled water and allow it to dry.
2. Cut the leaf disc of 4.5 cm diameter.
3. Prepare serial dilution of spore–crystal mixture in 0.2% Triton X-100.
4. Dip the leaf disc into different concentrations of spore–crystal mixture for 10 s and allow it to air-dry for 1 h.
5. Place the leaf discs into separate petri dishes. Use 0.2% Triton X-100 treated disc as a control.
6. Inoculate 6-day-old 8–10 larvae in each petri dish. A minimum of 40 larvae should be used for each concentration.
7. Allow larvae to feed for 48–72 h.
8. After each 24 h, record the larval mortality.
9. Estimate the LC_{50} on the basis of 48 h mortality data.

10. The concentration of spore–crystal mixture (μg/ml) required for 50% mortality (LC_{50}) can be obtained from plotting the logarithm of protein concentration against logarithm of mortal larvae.

4 Notes

- Use minimum six concentrations of spore–crystal mixture to graph a plot.
- At least five to six replicates should be kept for each concentration.
- The entire bioassay including control should be performed on the same day, at the same time to minimize the error.
- If control showing mortality of more than 20% than it should be discarded and experiment should be repeated again.

References

1. Mohan M, Gujar GT (2002) Geographical variation in larval susceptibility of the diamondback moth, *Plutella xylostella* (Lepidoptera: Plutellidae) to *Bacillus thuringiensis* spore–crystal mixtures and purified crystal proteins and associated resistance development in India. Bull Entomol Res 92:489–498
2. Herbert BN, Gould HJ, Chain EB (1971) Crystal Protein of *Bacillus thuringiensis* var. tolworthi: subunit structure and toxicity to *Pieris brassicae*. Eur J Biochem 24:366–375

Check for updates

Chapter 68

Purification and Characterization of Crystal Proteins

Abstract

For bioassay of crystal proteins and to study its biochemistry requires pure form of crystals. There are several techniques reported for crystal protein purification from *B. thuringiensis* including isopycnic centrifugation in cesium chloride, sodium bromide gradient centrifugation, centrifugation through step gradients of ludox, foam flotation-based extraction, carboxymethyl cellulose column chromatography, and gel filtration column chromatography. But the most widely used strategy is discontinuous sucrose density gradient ultracentrifugation. There is an intensive study going on related to the crystal protein endotoxin due to its insecticidal properties which may lead to its commercial production and its use as a biocontrol agent; therefore, there is a need to purify and characterize crystals proteins.

Keywords Crystal protein, Sucrose, Density gradient, Ultracentrifugation, Purification

1 Discontinuous Sucrose Density Gradient Ultracentrifugation Method

1.1 Principle

Sucrose density gradient ultracentrifugation is very commonly used technique for purifying biomolecules such as crystals from *Bacillus* species. This method works on the basis of density of the particles in the suspension. Particles that are denser than the solvent will be pelleted and those are less dense will float. In the spore–crystal mixture, spores are denser than crystals, so the spores will be found as pellet and less dense crystals will be float within the sucrose gradient. A high speed ultracentrifugation is used to accelerate the process in order to separate the spore and crystals to purify the crystals from the mixture [1].

2 Materials

- *GYS medium*: glucose 0.1 g, ammonium sulfate 0.2 g, yeast extract 0.2 g, K_2HPO_4 0.5 g, $MgSO_4{\cdot}7H_2O$ 0.02 g, $MnSO_4{\cdot}5H_2O$ 0.005 g, $CaCl_2$ 0.008 g, distilled water 100 mL, pH 7.2.
- Ice-cold sterilized distilled water.
- 50 mM Tris–HCl (pH 7.5).

M. Senthilkumar, N. Amaresan and A. Sankaranarayanan, *Plant-Microbe Interactions: Laboratory Techniques*, Springer Protocols Handbooks, https://doi.org/10.1007/978-1-0716-1080-0_68,

- Ultrapure sucrose.
- Centrifuge.
- Ultracentrifuge.

3 Methodology

3.1 Preparation of Spore–Crystal Mixture

1. Inoculate *B. thuringiensis* strain in GYS medium and incubate the broth on shaker at 150 rpm for 48 h at room temperature.
2. Harvest the completely autolysed spores and crystals by centrifugation at 4025 × *g* for 10 min at 4 °C.
3. Discard the supernatant and wash the pellet with ice-cold distilled water to remove cellular debris for 5–6 times.
4. View the spore–crystal mixture under phase contrast microscope to confirm the cellular debris free mixture.
5. Finally resuspend the pellet in 50 mM Tris–HCl (pH 7.5).
6. Separate the spores and crystals by using differential ultracentrifugation through a discontinuous sucrose density gradient.

3.2 Differential Discontinuous Sucrose Density Gradient Ultracentrifugation Method

1. Prepare 67%, 72%, and 79% (w/v) sucrose in 50 mM Tris–HCl (pH 7.5) containing 10 mM KCl.
2. Add 10 mL of 79% (w/v) sucrose at the bottom of the tube then add 10 mL of 72% and 67% (w/v).
3. Add 50 mg of spore–crystal mixture on the top of the 30 mL discontinuous sucrose gradient.
4. Carry out centrifugation in ultracentrifuge at 80,000 × *g* for 12 h at 4°C.
5. Crystal will form major band between 72% and 79% (w/v) sucrose while spores form discrete pellet at the bottom of the centrifuge tube.
6. Remove crystal band carefully with the use of the micropipette and wash it thrice with chilled 50 mM Tris–HCl (pH 7.5) by centrifugation at 15,000 × *g* for at 4 °C for 10 min.
7. Resuspend the final pellet in ice-cold distilled water and store the suspension at −20 °C for further use.
8. Purity of the crystals can be checked by phase-contrast microscopy and also by placing it on CCY agar plates after production of bacterial colonies.

4 Notes

- Take care of adding different sucrose concentrations; all the interfaces between each gradient should be clearly visible.

- After ultracentrifugation the tubes should be removed immediately taking care of not disturbing layers of the gradient.

Reference

1. Thomas WE, Ellar DJ (1983) *Bacillus thuringiensis* var *israelensis* crystal delta-endotoxin: effects on insect and mammalian cells in vitro and in vivo. J Cell Sci 60:181–197

Chapter 69

Purification of Crystal Proteins Using of Organic Solvent

Abstract

To purify the crystal protein, the protein is required in solution and it is possible by breaking the cells or tissues of the protein. There are various methods available for the purpose of purification and organic solvent extraction is one of them. The choice of the method employed depends on the sturdy of the cells and on its protein fragility. Through this method of organic solvent extraction, protein found in the solvent is the soluble protein which can be then separated from its DNA and cell membrane through the process of centrifugation.

Keywords Crystal protein, Purification, Organic solvent, Centrifugation

1 Principle

The above described methods are time consuming and require expensive instruments and reagent. The extraction using organic solvent is an inexpensive, rapid, and requires minimum equipment. Bt spores are hydrophobic in nature and crystals are of different shape, size of different Bt strain. So, by taking the advantage of this characteristic of spores, the method uses the organic solvent to trap the spores from spore–crystal mixture [1].

2 Materials

- Anderson medium.
- 1 M NaCl containing 0.01% Triton X-100.
- 10% aqueous solution of hexane.
- Ice-cold sterile distilled water.
- Saline solution.
- Sonicator.
- Centrifuge.

M. Senthilkumar, N. Amaresan and A. Sankaranarayanan, *Plant-Microbe Interactions: Laboratory Techniques*, Springer Protocols Handbooks, https://doi.org/10.1007/978-1-0716-1080-0_69,

3 Methodology

1. Grow *Bacillus* in Anderson medium for 48 h at room temperature on shaker at 150 rpm.
2. Centrifuge the completely autolysed culture taken in 50 mL centrifuge tube at 4025 × *g* for 10 min at 4 °C.
3. Wash the pellet by resuspending it in 1 M NaCl containing 0.01% Triton X-100. Repeat the procedure twice.
4. Suspend the pellet in saline solution.
5. Add 10% aqueous solution of hexane (organic solvent) to it.
6. Sonicate the suspension at 100 W for 10 min to disperse the clumps.
7. Centrifuge the suspension at 4025 × *g* for 10 min.
8. Resuspend the pellet again in saline solution, organic solvent, and repeat the same procedure thrice as above.
9. Spores will be trapped in hexane on the top layer while the crystals accumulate in the pellet and cell debris will be remained in aqueous phase.
10. Finally wash the pellet twice with ice-cold distilled water.
11. Air-dry the purified crystals and store it at −20 °C for further use.

4 Notes

- GYS broth can be used instead of Anderson medium.
- Use 50, 75, or 100 μL of aqueous solution of organic solvent (hexane) to prevent the altering of the crystals.

Reference

1. Mounsef JR, Salameh D, Kallassy AM, Chamy L, Brandam C, Lteif R (2014) A simple method for the separation of *Bacillus thuringiensis* spores and crystals. J Microbiol Methods 107:147–149

Check for updates

Chapter 70

Purification and Characterization of Vips from Culture Supernatant of *Bacillus*

Abstract

Bacillus thuringiensis, a well-known Gram-positive soil bacterium produces insecticidal proteins that accumulate in inclusion bodies during sporulation such as Cry and Cyt proteins. These inclusions are solubilized in insect midgut, releasing δ-endotoxins that exhibit insecticidal proteolytic activity. But there are several important insects such as lepidopteran black cutworm (BCW) that are less sensitive to their action. *B. thuringiensis* also produces proteins during their vegetative growth stage which are designated as VIP (Vegetative Insecticidal Protein) and are very effective against BCW. These second generation proteins are divided into four families Vip1, Vip2, Vip3, and Vip4 according to their amino acid sequence homology. Vip proteins from culture supernatant of *Bacillus* isolates can be purified by different methods and are screened for their insecticidal activity. The purification methods (1) anion exchange chromatography followed by SDS-PAGE gel electrophoresis and (2) Fast Protein Liquid Chromatography (FPLC) which are described here.

Keywords *Bacillus thuringiensis*, VIP (Vegetative Insecticidal Protein), Anion exchange chromatography, SDS-PAGE, Fast Protein Liquid Chromatography (FPLC)

1 Anion Exchange Chromatography Followed by SDS-PAGE Gel Electrophoresis

1.1 Principle

Commonly used technique for protein purification is anion exchange chromatography technique. In this, an impure extracted Vip protein is loaded into the ion exchange column at a particular pH. Charged protein will bind to the oppositely charged groups in resin. Proteins will be eluted with 250 mM NaCl and unwanted proteins and impurities will be removed by washing the column [1, 2].

1.2 Materials

- *Terrific broth*: 12% tryptone, 2.4% yeast extract, 0.04% glycerol, 0.17 M KH_2PO_4, 0.72 M K_2HPO_4.
- Ammonium sulfate.
- 20 mM Tris–HCl buffer (pH 7.5).
- 20 mM sodium citrate buffer (pH 2.5).
- *Buffer and reagents for anion exchange chromatography*: 20 mM Bis Tris-Propane (pH 9.0), 500 mM NaCl, 250 mM NaCl.

M. Senthilkumar, N. Amaresan and A. Sankaranarayanan, *Plant-Microbe Interactions: Laboratory Techniques*, Springer Protocols Handbooks, https://doi.org/10.1007/978-1-0716-1080-0_70, © Springer Science+Business Media, LLC, part of Springer Nature 2021

- *Reagents for SDS-PAGE gel electrophoresis*: 40% acrylamide solution, 1% bisacrylmide solution, 1.5 M Tris–HCl (pH 8.7), 10% APS, 10% SDS, TEMED, silver stain.
- Centrifuge.
- Poros HQ/N anionic exchanger.
- SDS-PAGE gel electrophoretic unit.

1.3 Methodology

1. Grow the *Bacillus* species in Terrific broth at 30 °C for 16–20 h.
2. After incubation, centrifuge the cultures at 16,099 × *g* for 10 min at 4 °C.
3. Collect the supernatant containing Vip proteins and follow the ammonium sulfate precipitation method to precipitate protein.
4. Add slowly solid $(NH_4)_2SO_4$ to a bacterial supernatant at 60% saturation by mixing on a magnetic stirrer at 4 °C.
5. Centrifuge the precipitated material at 11,180 × *g* for 10 min at 4 °C.
6. Resolve the pellet by adding 1 mL of 20 mM Tris–HCl buffer (pH 7.5).
7. To remove the salts and small molecules, dialyze the material overnight against 20 mM Tris–HCl buffer (pH 7.5).
8. After overnight dialysis at 4 °C, titrate the dialyzate to pH 4.5 using 20 mM sodium citrate (pH 2.5).
9. Incubate it for 30 min at room temperature.
10. Then centrifuge it at 2795 × *g* for 10 min.
11. Dissolve the resulting pellet in 20 mM Bis Tris-Propane (pH 9.0) and fractionate it on anion exchange column using Poros HQ/N anionic exchanger.

1.3.1 Anion-Exchange Chromatography

1. Use 500 mM NaCl in 20 mM Bis Tris-Propane (pH 9.0) at a flow rate of 4 mL/min to equilibrate the column.
2. Load the sample and elute insecticidal protein with 250 mM NaCl.
3. Wash the column with buffer.

1.3.2 SDS-PAGE Gel Electrophoresis

1. To determine the purity of the eluted protein sample, run the obtained fractions on 12% SDS-PAGE.
2. Stain the gel with fast silver staining technique.
3. Scan the gel image using computer.

1.4 Notes

- Use vegetative phase of growth.
- Use freshly prepared solutions.

- Q-Sepharose (anion exchanger) can also be used instead of Poros HQ/N anionic exchanger.

2 Fast Protein Liquid Chromatography (FPLC)

2.1 Principle

Vegetative insecticidal proteins can be purified by the fast protein liquid chromatography (FPLC) technique in which the mobile phase is an aqueous buffer and the flow rate is kept constant by a positive-displacement pump. The stationary phase is composed of a bead (usually of cross-linked agarose) which is packed into a cylindrical column. The eluent is passed through the detectors to measure the protein concentration by absorbing the UV light at 280 nm [3, 4].

2.2 Materials

- LB broth.
- 20 mM Tris–HCl buffer (pH 7.5).
- Sonicator.
- Ammonium sulfate.
- 50 mM Tris–HCl (pH 7.5).
- 1× PBS.
- Centrifuge.

For FPLC

- Superdex 75 gel filtration column.
- Buffers A (10 mM Tris–HCl, pH 7.0).
- Buffer B (10 mM Tris–HCl, pH 7.0, 1.0 M NaCl).

2.3 Methodology

1. Grow the *Bacillus* species on LB medium for 48 h at 28 °C.
2. After incubation, centrifuge at 4500 × g for 10 min, discard the supernatant and dissolve the pellet in 20 mM Tris–HCl buffer (pH 7.5).
3. Sonicate it on ice for 30 s at 900 psi pressure using a Spectronics instrument that is equipped with a mini-cell. The flow rate should be 18–20 drops per min.
4. Centrifuge the supernatant at 13,000 × *g* for 10 min.
5. Precipitate the protein slowly by adding ammonium sulfate to a 40% saturation level for 3 h at 4 °C.
6. Collect the precipitated protein by centrifugation at 13,000 × *g* for 15 min.
7. Dissolve the pellet in 50 mM Tris–HCl (pH 7.5) and then dialyzed overnight with 1× PBS buffer.

8. Purify the Vips by FPLC using gel filtration Superdex 75 column.

2.3.1 Ion-Exchange FPLC Protocol

1. Prime the pumps A and B with degassed buffers A (10 mM Tris–HCl, pH 7.0) and buffer B (10 mM Tris–HCl, pH 7.0, 1.0 M NaCl), respectively.
2. Equilibrate the Superdex 75 column with 5 volumes of buffer A and 10 volumes of buffer B then again with 5 volumes of buffer A.
3. Wash the sample loading loop with buffer A.
4. Load the sample protein 500 μg/mL into the column.
5. Monitor the eluted protein by UV absorbance at 280 nm wavelength.

2.4 Notes

- Use fresh culture of the organism which is in vegetative phase. Do not use very old sporulated culture.
- Use freshly prepared buffer solutions.

References

1. Estruch JJ, Warren GW, Mullins MA, Nye GJ, Craig JA, Koziel MG (1996) Vip3A, a novel *Bacillus thuringiensis* vegetative insecticidal protein with a wide spectrum of activities against lepidopteran insects. Proc Natl Acad Sci U S A 93:5389–5394
2. Chakroun M, Banyuls N, Bel Y, Escriche B, Ferré J (2016) Bacterial vegetative insecticidal proteins (Vip) from entomopathogenic bacteria. Microbiol Mol Biol Rev 80:329–350
3. Güney E, Adıgüzel A, Demirbağ Z, Sezen K (2019) *Bacillus thuringiensis kurstaki* strains produce vegetative insecticidal proteins (Vip 3) with high potential. Egypt J Biol Pest Control 29:81
4. Osman G, Assaeedi A, Osman Y, El-Ghareeb D, Alreedy R (2013) Purification and characterization of *Bacillus thuringiensis* vegetative insecticidal toxin protein(s). Lett Appl Microbiol 57:310–316

Chapter 71

Detection of Vegetative Insecticidal Protein (Vip) in *Bacillus*

Abstract

Vegetative insecticidal protein in *Bacillus* can be detected through the process of PCR (Polymerase Chain Reaction). PCR approach is based on the amplification of the gene and will assign the amplified gene into the Vip gene family but this approach will not identify the specific gene of the family. Therefore, PCR-RFLP technique is used to detect the type of the Vip genes within a same family.

Keywords Vegetative insecticidal protein, Polymerase chain reaction, PCR-RFLP, *Bacillus*

1 Principle

Bacillus thuringiensis and *B. cereus* secretes some toxic proteins during vegetative growth that do not form crystals called Vip proteins which does not have homology with Cry proteins. Vip proteins bind to different target sites than Cry proteins. Vip proteins together with Cry proteins can be used in insect pest management. This combination of toxins could be applied to widen the spectrum of toxicity and to minimize the risk of cross-resistance. Vip genes can be identified or detected by polymerase chain reaction-restriction fragment length polymorphism (PCR-RFLP). Detection of Vip1, Vip2, and Vip3 genes in strains of *B. thuringiensis* can be performed by using polymerase chain reaction (PCR)-based methods. Amplification products of PCR reactions can confirm the presence of Vip genes, but this approach only assigns the gene into a Vip family (Vip1, Vip2, or Vip3) and does not allow the identification of specific genes within a family. Restriction fragment length polymorphism (RFLP) was therefore used to identify the different types of Vip genes within a same family [1].

2 Materials

- *Bacillus thuringiensis* strains.
- Taq polymerase.

M. Senthilkumar, N. Amaresan and A. Sankaranarayanan, *Plant-Microbe Interactions: Laboratory Techniques*, Springer Protocols Handbooks, https://doi.org/10.1007/978-1-0716-1080-0_71,

- 10× reaction buffer.
- Deoxynucleotide triphosphate.
- Primers for Vip gene.
- Thermocycler.
- Gel electrophoretic unit.

3 Methodology

3.1 Screening of Vip1, Vip2, and Vip3 Genes by PCR

- Extract the DNA by any standard protocol.
- Perform the screening of Vip genes with primers designed from conserved regions within the Vip gene families.
- Make a final volume of 25 μL PCR mixture which includes 1 μL of the DNA template, 1 U of Taq polymerase, 2.5 μL of 10× reaction buffer, 250 mmol/L (each) of deoxynucleotide triphosphate, and primers for Vip gene screening (50 pmol/L for Vip1, Vip2, and Vip3 primers) as mentioned in table.
- Carry out the amplifications in a thermal cycler with 5 min denaturation at 95 °C followed by 35 cycles of amplification with 1 min denaturation at 95 °C, 1 min of annealing at 45–47 °C, and 1.5 min of extension at 72 °C and final extension step of 10 min at 72 °C.

3.2 Identification of vip Gene Types by PCR-RFLP

- Perform a second PCR with "typing" primers for the further identification of Vip1 and Vip2 family genes as per given in table.
- Perform PCR reactions as mentioned above primers using 50 pmol/L of Vip1- and Vip2-typing primers.
- Carry out the PCR cycling for these reactions as follows: 5 min denaturation at 95 °C followed by 35 cycles of amplification with 1 min denaturation at 95 °C, 1 min of annealing at 45–48 °C and 1.5 min of extension at 72 °C, and final extension step of 10 min at 72 °C.
- Digest the PCR fragment from Vip1 and Vip2 "typing" amplification, and from Vip3 "screening" amplification, with *AluI* and *Sau3AI* separately.
- Prepare the digest mixture as follows: 5 μL of PCR product, 5 U of the enzyme and 2 μL of 10× appropriate reaction buffer in a total volume of 20 μL.
- Incubate the digest mixture at 37 °C for 4 h and analyze the digestion products by agarose gel electrophoresis.

Primers used for screening and typing Vip genes

Primer	Sequence[a]	Product size at (bp)
Screening		
Vip1-sc.fw	5′-TATTAGATAAACAACAACAAGAATATCAATCTATTMG NTGGATHGG-3′	585
Vip1-sc.rev	5′-GATCTATATCTCTAGCTGCTTTTTCATAATCTSARTAN GGRTC-3′	
Vip2-sc.fw	5′-GATAAAGAAAAAGCAAAAGAATGGGRNAARRA-3′	845
Vip2-sc.rev	5′-CCACACCATCTATATACAGTAATATTTTCTGGDATNGG-3′	
Vip3-sc.fw	5′-TGCCACTGGTATCAARGA- 3′	1621
Vip3-sc.rev	5′-TCCTCCTGTATGATCTACATATGCATTYTTRTTRTT-3′	
Typing		
Vip1-tp.fw	5′-AAACGGGTGATTTYACNTT-3′	1044
Vip1-tp.rev	5′-GGGAATTAAAATCATCCAT-3′	
Vip2-tp.fw	5′-AGAATGGGGGAAAGARAA-3′	1107
Vip2-tp.rev	5′-ACCTCTGTTACTTTATCDAT-3′	

[a]Universal code for degenerate bases: R = A, G; Y = C, T; M = A, C; S = G, C; H = A, T, C; D = G, A, T; N = A, C, G, T

4 Note

- PCR approach will allow the assignment of the amplified gene into a Vip gene family, but it will not allow identification beyond this level. Therefore, a PCR-RFLP technique can be used to detect Vip genes and to classify them within their types.

Reference

1. Hernández-Rodríguez CS, Boets A, Van Rie J, Ferré J (2009) Screening and identification of vip genes in *Bacillus thuringiensis* strains. J Appl Microbiol 107:219–225

Check for updates

Chapter 72

Selective Isolation of Entomopathogenic Fungi

Abstract

Entomopathogenic fungi kill insects; arachnids and fungi regulates the population of the host. In agricultural sector, entomopathogenic fungi are used to control the pest population, therefore, can be used as a biocontrol agent. During the traditional times, biocontrol activity of entomopathogenic fungi is shown through the application of fungi material i.e., conidia to the agricultural crops by inoculative biocontrol technique but the drawback of this approach is the exploitation of the indigenous fungi reservoir of the previous cropping system. Hence, a modern approach is used to isolate entomopathogenic fungi by making use of selective media.

Keywords Entomopathogenic fungi, Selective media, Biological control, *Metarhizium anisopliae*, *Beauveria bassiana*

1 Principle

Entomopathogenic fungi are used for biological control of arthropods, and these are worldwide. The successful examples of fungus-based insect control programs includes *Metarhizium anisopliae* sensu lato (s.l.) (Metschnikoff) Sorokin and *Beauveria bassiana* (Balsamo) *Vuillemin* are the fungi most commonly employed for pest control. Two *M. acridum* (one African and one Australian) have been developed as commercial products for biological control of the insect group of locusts and grasshoppers (Acrididae). To isolate the entomopathogenic fungi from soil, two most commonly employed methods are: (1) baiting the environment with a susceptible insect host i.e., Galleria Bait technique or (2) use of specific selective media containing chemicals that reduce the growth of contaminants. For the isolation of a specific fungus, a selective media should contain both a nutrient source and antimicrobial agents (i.e., fungicides and antibiotics), in appropriate concentrations to grow the target fungus and to avoid the growth of common contaminants.

M. Senthilkumar, N. Amaresan and A. Sankaranarayanan, *Plant-Microbe Interactions: Laboratory Techniques*, Springer Protocols Handbooks, https://doi.org/10.1007/978-1-0716-1080-0_72, © Springer Science+Business Media, LLC, part of Springer Nature 2021

2 Materials

- Potato dextrose agar.
- Yeast extract.
- Gentamicin.
- *N*-dodecylguanidine monoacetate.
- Fungicide.
- Chloramphenicol.
- Thiabendazole.
- Cycloheximide.
- Oatmeal agar.
- Chlortetracycline.
- Dodine.
- Deionized water.
- Glucose.
- Oxgall.
- Peptone.
- Rose bengal.
- Cheesecloth.
- Mortar and pestle.

3 Methodology

Prepare the selective media as follows:

- **Dodine medium**: PDAY (potato dextrose agar supplemented with 1 g/L yeast extract), +0.05 g/L gentamicin, and one of four dodine (*N*-dodecylguanidine monoacetate) concentrations (0.001%, 0.002%, 0.004% or 0.006% active ingredient (A.I.)). The commercial fungicide "Syllit" (65% A.I.) can be used as the dodine source. Autoclave an aqueous stock solution of dodine (0.1% A.I.) at 121 °C for 15 min separately and then mix thoroughly with an autoclaved PDAY in appropriate quantities to obtain the required concentration [1].
- **CTC medium**: Prepare PDAY supplemented with 0.5 g/L chloramphenicol, 0.001 g/L thiabendazole and 0.25 g/L cycloheximide. Autoclave as per standard procedure. Also one can prepare CTC medium with higher concentrations of thiabendazole, i.e., 0.002 and 0.004 g/L. CTC media with these three thiabendazole concentrations can be labeled as "'CTC1T,"

"CTC 2T," and "CTC 4T." "CTC" without thiabendazole concentration is equivalent to "CTC 1T." [1]

- **Oatmeal agar medium:** Prepare 20 g oatmeal agar, autoclave it for 20 min with 1 L of deionized water. Filter the suspension immediately twice through four layers of cheesecloth, and adjust the volume of filtrate to 1 L with deionized water. Grind the agar (20 g/L) in a mortar and pestle with various amounts of dodine and then add it to the oatmeal filtrate by stirring. Autoclave the final medium for 20 min and cool it to 48 °C prior to adding chlortetracycline (5 mg/L) [2].
- **Veen's medium**: Veen's medium contains glucose, oxgall, and peptone with rose bengal, chloramphenicol, and cycloheximide as antibiotics. Adjust the pH of all media to 6.9 with 10% NaOH solution, before autoclaving at 121 °C for 15 min, and pipette out 23 mL of each medium individually into a polystyrene petri plate (95 × 15 mm). Store the plates of media at room temperature in the dark and use them within 4 days of preparation [2].

4 Notes

- **Dodine medium** will allow the growth of *M. brunneum*, *M. robertsii*, *B. bassiana*, and *B. brongniartii* from non-sterile soil but not the *M. acridum* isolates because *M. acridum* is very susceptible to dodine even at very low concentration; therefore, CTC medium is used for isolation of *M. acridum*.
- **CTC medium** is especially used for the isolation of *M. acridum* fungi.
- **Oatmeal agar medium** can be used for isolating *Metarhizium anisopliae and Beauveria bassiana* from an artificial potting medium.
- **Veen's medium** is designed to maximize recovery of naturally occurring *Beauveria* spp. and *Metarhizium* spp., especially *M. acridum*, from soil.

References

1. Fernandes ÉK, Keyser CA, Rangel DE, Foster RN, Roberts DW (2010) CTC medium: a novel dodine-free selective medium for isolating entomopathogenic fungi, especially *Metarhizium acridum*, from soil. Biol Control 54:197–205
2. Chase AR, Osborne LS, Ferguson VM (1986) Selective isolation of the entomopathogenic fungi *Beauveria bassiana* and *Metarhizium anisopliae* from an artificial potting medium. Fla Entomol 69:285–292

Chapter 73

Galleria Bait Technique for the Isolation of Entomopathogenic Fungi from Soil

Abstract

Isolation of entomopathogenic fungi using selective media manipulates the saprotrophic ability of the hypocreale an entomopathogenic fungi. To manipulate the fungi ability to infect host, Galleria Bait Technique is used. Earlier, this technique is used to isolate entomopathogenic nematodes from various soil samples and was rarely used for isolating fungi. Then in 1986, Zimmermann declared this technique as a standard protocol for isolating entomopathogenic fungi.

Keywords *G. mellonella*, Entomopathogenic, Galleria Bait, Isolation, *Ixodes scapularis*

1 Principle

Entomopathogenic fungi are mainly found in forest soils and can be isolated from the interface between leaf litter (Oi horizon) and the organic (Oe + Oa horizons) layer of soil. *Ixodes scapularis*, the black-legged tick, is an ectoparasite and a vector for several disease-causing agents. Presently, use of pesticides is the most effective method to control the tick. But due to increasing resistance of ticks to pesticides and because of the negative effects of the pesticides application on the environment, application of entomopathogenic fungi to soil surfaces could be used as a biological control agent for controlling tick populations. Therefore, the *Galleria* bait method is an efficient method for isolation of these fungi from soils [1].

2 Materials

- Soil samples.
- Plastic containers.
- *Galleria mellonella* L. (greater wax moth) larvae.
- Known entomopathogenic fungal cultures.
- 1% sodium hypochlorite solution

M. Senthilkumar, N. Amaresan and A. Sankaranarayanan, *Plant-Microbe Interactions: Laboratory Techniques*, Springer Protocols Handbooks, https://doi.org/10.1007/978-1-0716-1080-0_73,

- Penicillin–streptomycin solution.
- Distilled water.

3 Methodology

1. Isolate the entomopathogenic fungi from soil using *Galleria mellonella* L. (greater wax moth) larvae.
2. The soil samples are maintained at 4 °C until they used in the assay.
3. Place the *G. mellonella* larvae in a plastic container of each of the soil samples; seal the containers with perforated lids and place them at room temperature.
4. Place three to five *G. mellonella* larvae in containers with sterile soil (negative control), no soil (negative control), or sterilized soil to which fungi obtained from one plate (approximate colony surface area of 28.3 cm^2) of each of the known entomopathogenic fungal cultures added (positive controls).
5. Observe the containers every other day, and collect the dead larvae.
6. Surface sterilize the carcasses for 3 min in a 1% sodium hypochlorite solution, rinse it in sterile distilled water, perform plating, and incubate at 27 °C in a humidity chamber at 100% relative humidity to permit growth of fungi.
7. Dissect the larvae and plate them on half-strength PDA having penicillin–streptomycin solution.
8. Record the number of plates infected with fungus, the number of entomopathogenic fungal colonies obtained, and also note down the number of fungal colonies suspected to be entomopathogenic.

4 Notes

- Dilute all samples in a ratio of 1:100 or 1:1000 in a potassium phosphate buffer (232 mM) containing 0.2 M monobasic potassium phosphate and 0.4 M dibasic potassium phosphate and plate them on potato dextrose agar (PDA) media to assess the frequency of entomopathogenic fungi found in samples.
- Prepare the half-strength PDA plates with a 10 mg/L penicillin-streptomycin solution (1663 U Penicillin G/mg, 748 U streptomycin sulfate/mg).

Reference

1. Tuininga AR, Miller JL, Morath SU, Daniels TJ, Falco RC, Marchese M, Stafford KC (2014) Isolation of entomopathogenic fungi from soils and *Ixodes scapularis* (Acari: Ixodidae) ticks: prevalence and methods. J Med Entomol 46:557–565

Check for updates

Chapter 74

Microscopic Characterization of Entomopathogenic Fungi

Abstract

The most widely used criteria for characterization of entomopathogenic fungi is its microscopic characterization of asexual morphological characters. It requires adequate observation of both conidia and conidiogenous cells. There are number of methods used for microscopic observation. Usually whole mount method is used for observing fungi under light microscope. If done correctly, it is very easy and rapid method but disruption of conidiogenous cells and dehiscence of conidia is also very common while preparing slide. Therefore, another method is routinely used for more critical observation of conidiogenesis known as slide culture preparation. The problem of disruption of conidiogenous cells can easily be avoided with this method but the culture must be incubated long enough to develop conidiogenesis for observation.

Keywords Entomopathogenic fungi, Light microscope, Conidia, Scanning electron microscopy, Light microscopy

The most widely used criteria for characterization of entomopathogenic fungi is its microscopic characterization of asexual morphological characters. It requires adequate observation of both conidia and conidiogenous cells. There are number of methods used for microscopic observation. Usually whole mount method is used for observing fungi under light microscope. If done correctly, it is very easy and rapid method but disruption of conidiogenous cells and dehiscence of conidia is also very common while preparing slide. Therefore, another method is routinely used for more critical observation of conidiogenesis known as slide culture preparation. The problem of disruption of conidiogenous cells can easily be avoided with this method but the culture must be incubated long enough to develop conidiogenesis for observation.

M. Senthilkumar, N. Amaresan and A. Sankaranarayanan, *Plant-Microbe Interactions: Laboratory Techniques*, Springer Protocols Handbooks, https://doi.org/10.1007/978-1-0716-1080-0_74, © Springer Science+Business Media, LLC, part of Springer Nature 2021

1 Light Microscopy

1.1 Materials

- PDA agar plate.
- Water agar plate.
- **Mounting medium**
  - Lactophenol cotton blue: lactic acid 80 mL, phenol crystals 100 g, glycerin 159 mL, distilled water 100 mL, mix all the ingredients and heat it. Add 0.5% cotton blue.
  - Lacto-aceto-orcein solution: 45% glacial acetic acid 50 mL, 85% lactic acid 50 mL, orcein 2 g. Mix ingredients in flask and heat until hot, do not boil. Hold for 30 min and filter it while hot. Cool and dilute it to 1:3 with 45% acetic acid.
- Sterile glass coverslip.
- Glass slides.
- Insect pin or tooth pin.
- Light microscope.

1.2 Methodology

1.2.1 Whole Mount Method [1]

1. Grow fungus on PDA plate and allow it to grow at 28–30 °C until sporulation.
2. After appropriate incubation, place fungal mycelium on coverslip.
3. Place a drop of mounting medium and tease the growth carefully.
4. Place the second coverslip on it and place coverslip sandwich on microscopic slide.
5. Observe under 40× of light microscope.

1.3 Notes

- Use glass coverslip, do not use plastic coverslip. Plastic coverslips are thick and little hazy than the glass coverslip.
- Use minimum amount of mounting medium. Too much mounting medium will float the coverslip.

1.3.1 Slide Culture Method

1. Cut 1.0 cm agar block from previously prepared agar plate.
2. Put it on a sterile coverslip and place the coverslip on 2% water agar plate.
3. Inoculate the agar block with respective fungal tissue.
4. Place a second sterile coverslip on the top of the inoculated block and cover the petri dish with the lid.
5. Incubate the petri dish at 28–30 °C for few days until the visible growth appears.
6. Once the fungus has started sporulation, remove the agar block from the coverslip and mount the coverslip on the slide with

appropriate mounting medium and observe under the light microscope at 40×.

1.4 Notes

- Culture must be incubated for a long period of time to allow the conidiogenesis to observe the whole conidiogenous cell.
- Usually 3–4 replicate cultures should be recommended to establish to face a situation where sporulation may not be observed within an incubation time.

2 Scanning Electron Microscopy

2.1 Materials

1. Potato dextrose broth.
2. **Working solutions**
   (a) 0.2 M Cacodylate buffer: cacodylate buffer stock 50 mL, 0.2 M HCl 6.0 mL, make up to 100 mL with double distilled water.
   (b) 2.5% Glutaraldehyde: 8.0% glutaraldehyde 10 mL, 0.2 M cacodylate buffer 16 mL, double distilled water 6.0 mL.
   (c) 1% osmium tetroxide: 4% osmium tetroxide 1 mL, 0.3 M sucrose 1 mL, 0.2 M Cacodylate buffer 2 mL (wrap the vial in foil).
3. Acetone solutions: 45%, 60%, 75%, 90% 100%.
4. Hexamethylene disilazane (HMDS).
5. Silica gel.
6. Aluminum foil or silver stub.
7. Gold particles.
8. Cathodic sprayer.
9. Critical point dryer.
10. Desiccator.
11. Scanning electron microscope.

2.2 Methodology [2]

1. Grow entomopathogenic fungus of interest in potato dextrose broth for 2–4 days at 28–30 °C.
2. Take 10 μL of fungal culture in micro-centrifuge tube and add 2.5% Glutaraldehyde fixative for 2 h.
3. Then wash it three times with sterile distilled water for 10 min each time.
4. Post-fix in 1% osmium tetroxide solution for 2 h.
5. Rinse three times with distilled water.
6. Follow dehydration in serially diluted (45%, 60%, 75%, 90%, and 100%) acetone solutions.

7. Transfer sample to the critical point dryer to completely dry with hexamethylene disilazane (HMDS).
8. Mount the sample on aluminum foil or silver stub and cover with gold with cathodic spraying.
9. Keep the material in desiccator with silica gel until observing.
10. Observe the sample under SEM.

2.3 Notes

- Small piece of agar block with the fungal growth can also be used instead of liquid culture.
- HMDS can be replaced by carbon dioxide in critical point dryer.

References

1. Mishra S, Kumar P, Malik A (2015) Morpho-molecular characterization and virulence determination of entomopathogenic fungal isolates native to Indian subcontinent. Med Mycol 1:3
2. Alves E, Lucas GC, Pozza EA, de Carvalho Alves M (2013) Scanning electron microscopy for fungal sample examination. In: Laboratory protocols in fungal biology. Springer, New York, pp 133–150

Chapter 75

Isolation and Characterization of Nematode Species with Economic Importance

Abstract

The economically important nematodes known as entomopathogenic nematodes (EPN) act as a biocontrol agent against various insects pests. EPN generally belong to the families Steinernematidae and Heterorhabditidae. Economically important nematodes are considered as nutrient recycler as it provides the essential nutrients to improve the growth of the plant. EPN also helps in decomposition of organic matter, so various studies are focused to isolate and characterize economically important nematodes by methods like Baiting technique, Baermann funnel method, etc.

Keywords Entomopathogenic nematodes (EPN), Nutrient recycler, Baiting technique, Baermann funnel

1 Baiting Technique

1.1 Principle

Nematodes are microscopic worms present in soil and water. Some nematodes are beneficial and some are harmful to the plants. The beneficial nematodes control the grubs and other insects in the soil and harmful nematodes that damage the plants by burrowing in to root, stems, and leaves. Plant damage makes it difficult to absorb nutrients, resulting in weak, small, or dead flowering [1].

1.2 Materials

- Soil sample.
- Petri dish.
- 150-mesh metal screen.
- Larvae.
- Filter paper.

1.3 Methodology

- A nematode trap place on the 0.5 kg soil sample.
- A nematode trap consists of 5-cm-diameter petri dish covered with a 150-mesh metal screen and containing four last instar *Galleria mellonella* larvae.

M. Senthilkumar, N. Amaresan and A. Sankaranarayanan, *Plant-Microbe Interactions: Laboratory Techniques*, Springer Protocols Handbooks, https://doi.org/10.1007/978-1-0716-1080-0_75, © Springer Science+Business Media, LLC, part of Springer Nature 2021

- Two traps are buried in each soil sample and incubate at 25 °C for 6–7 days and remove the larvae.
- Place dead larvae on the moist filter paper in 5-cm-diameter plastic petri dish cover, which is glued, open side up to the center of a large petri dish (9-cm-diameter).
- Fill the petri plate with 3 mm of water (3 mm water-0.12 in.).
- Incubate the dead larvae in this system at 25 °C for another 6–7 days.
- Collect the emerging nematodes infective stage from the water suspension from the petri dish.
- Collected nematodes from dish are re-exposed to *Galleria mellonella* larvae in 5-cm-diameter petri dish with moist filter paper.
- Only the newly emerging infective juveniles' are used for further study.
- Identify nematodes by microscopic technique.

2 Baermann Funnel Method

2.1 Principle

Baermann funnel method is the classical method used to isolate different types of nematode species and useful for specific isolation of nematodes that are not attracted by *E. coli* or without *E. coli* contamination [2].

2.2 Materials

- Baermann funnel.
- Rubber tube.
- M9 solution (Dissolve the salt in 800 mL water and adjust the pH to 7.2 with NaOH. Make the final volume of 1 L and autoclave it).
- NaOH (for pH adjustment).

2.3 Methodology

- Take Baermann funnel and fit a rubber tube on to the stem of it.
- Close the rubber tube with clip and fill the rubber tube with M9 solution up to the neck of the funnel.
- Cover the funnel cone with a piece of sieve then fill with sample and fold to enclose the sample.
- Wet the sample by adding M9 solution or water.
- Active nematodes crawl down through the sieve at the end of the rubber tube.
- After few hours recover the nematodes by opening the clamp.

2.4 Notes

- Sieve piece could be standard cleaning tissue, muslin cloth, mosquito mesh.
- The Baermann funnel method is not quantitatively reliable because it is based on against slow and non-moving nematodes.

3 Centrifugal Flotation

3.1 Principle

Centrifugal flotation method is a well-suited method to recover both motile and nonmotile nematodes. Nematodes are separated from the macerated sample by centrifugal flotation using a solution of specific weight (higher than the nematodes). The solution could be sucrose, $MgSO_4$, and Ludox, so the worms will float and soil particles will sink. The sinking fraction and floating fraction can be separated by centrifugation [3].

3.2 Materials

- Soil sample.
- Centrifuge tube.
- Density solution ($MgSO_4$, sucrose, Ludox-colloidal silica).
- M9 solution.
- NGM culture media plate.

3.3 Methodology

1. Collect soil sample and wash with 1 mm mesh to remove the debris (if soil sample contain lot of debris).
2. Soak the sample in water for brief time and then centrifuge the sample at 1800 g to obtain the pellet and discard the supernatant.
3. Resuspend the pellet in 20 mL of the solution (according to your choice) of density and centrifuge at 1800 rpm for 1 min.
4. After centrifuge add 2 mL of water or M9 solution on top and centrifuge again for 5–15 min at 1800 rpm (nematodes will accumulate at the interface between two solutions).
5. Collect the nematodes using a cut 1 mL micropipette tip and rinse them in M9 buffer, again centrifuge, and pipette the pellet on a standard NGM culture plate.

3.4 Notes

- NGM is a Nematodes Growth Medium. It is used for maintaining nematodes.
- Sucrose: It is inexpensive, sticky, has strong osmotic effect that harms the worms (1.71 mol/L for a density of 1.15).
- $MgSO_4$: It is less sticky than sucrose and also has strong osmotic effect (1.38 mol/L).

- Ludox-colloidal silica: this is expensive, weak osmotic effect (1.38 mol/L).

4 Characterization of Nematodes

4.1 Morphological Characterization [4]

4.1.1 Materials

- TAF fixative (14 mL of 40%, of formaldehyde + 4 mL triethanolamine in 82 mL distilled water).
- 95% ethanol.
- Glycerin.
- Distilled water.
- Compound or digital microscope.

4.1.2 Methodology

1. Heat-kill the collected nematodes in a water bath at 50–60 °C for 3 min.
2. Fix the specimens in 2–3 drops of double strength.
3. Use the temporary mounts within a week of preparation.
4. Place the fixed nematodes in an excavated glass block and 7 mL of solution 1 (20 parts of 95% ethanol and 1 part of glycerin in 79 parts of distilled water).
5. The glass blocks with samples are slowly dried by evaporation in a desiccator stuffed with cotton soaked in 95% alcohol.
6. Place the glass block in an oven at 35 °C for 12 h.
7. After 12 h flood with Solution 2 (95 parts of 95% ethanol and 5 parts of glycerin); put in the oven at 40 °C for 4 h.
8. Mount the nematodes on microscope slides. Cover the mounts with cover slip supported with wax or nail vanish.
9. Observe the morphological features such as nematode body length (L), maximum body diameter (MBD), anal body diameter (ABD), length tail (TL), distance from anterior end to esophagus/pharynx (ES), spun from anterior end to excretory pore (EP), hyaline (H) and in males spicule length (SL), gubernaculum length (GL), presence and absence of mucron.
10. Express the morphometric in micrometers.
11. Perform the experiment in multiple times, calculate means and ratios. The collected results can be used for comparison with existing information of described nematodes.

4.2 Molecular Identification

4.2.1 18S rDNA Sequencing

The identification of nematodes can be carried out by sequencing of the 18S ribosomal RNA gene sequencing with nematode specific primers [4, 5].

1. SSU18A (5′-AAAGATTAAGCCATGCATG-3′) and SSUR (5-′-CATTATTGGCAAATGATTTCG-3′) which amplify fragment about 1 kb of 18S rDNA in rhabditid nematodes.
2. RHAB1350F (5′-TACAATGGAAGGCAGCAGGC-3′) and RHAB1868R (5′-CCTCTGACTTTCGTTCTTGATTAA-3′) which amplify fragment of about 500 bp.
3. G18S4a (5′-GCTCAAGTAAAAGATTAAGCCATGC-3′) and DF18S (5′-GTTTACGGTCAGAACTASGGCGG-3′), which amplify the fragment around 1 kb in Rhabditidae, Cephalobidae, and Panagrolaimidae.

Follow the standard PCR protocol for amplification of 18S rDNA and proceed for sequencing.

References

1. Glazer I, Liran N, Steinberger Y (1991) A survey of entomopathogenic nematodes (Rhabditida) in the Negev desert. Phytoparasitica 19:291–300
2. Salame L, Glazer I, Miqaia N, Chkhubianishvili T (2010) Characterization of populations of entomopathogenic nematodes isolated at diverse sites across Israel. Phytoparasitica 38:39–52
3. Barrière A, Félix MA (2006) Isolation of *C. elegans* and related nematodes. WormBook 17:1–9
4. Ngugi CN, Haukeland S, Wachira PM, Mbaka JN, Okoth S (2019) Morphometrics and morphological characterization of entomopathogenic nematode isolate TK1 from Kenya. Int J Agric Environ Biores 4:80–90
5. Tian ZL, Maria M, Barsalote EM, Castillo P, Zheng JW (2018) Morphological and molecular characterization of the rice root-knot nematode, *Meloidogyne graminicola*, Golden and Birchfeild, 1965 occurring in Zhejiang, China. J Integr Agric 17:2724–2733

Chapter 76

Isolation of Nematodes from Soil Sample

Abstract

The most efficient and easy method to isolate nematodes from soil sample is to isolate on agar culture plate for fast moving nematodes. For isolation of active nematodes from soil, substrate, or sediment, the decanting method is followed. This method is based on the specific gravity difference between nematodes and soil particles and is also based on the mobility of the nematodes.

Keywords Nematodes, Agar culture method, Decanting method, Mobility

1 Agar Culture Plate Technique

1.1 Principle

Agar culture plate technique is a simple method to isolate fast moving and bacteria feeding nematodes from soil as well as compost sample. By using this method all developmental stages can be recovered and the number of individuals in soil sample can be measured. There are some nematode species associated with invertebrates during the dauer stage. These dauer stage nematodes can be recovered by scarifying the carriers placing on NGM plates [1].

1.2 Materials

- Soil sample.
- *E. coli* OP50 culture.
- NGM media plate.
- M9 solution or water.

1.3 Methodology

- Prepare NGM plates with *E. coli* OP50 lawn growth (1 cm).
- Spread 1–2 g of soil around the *E. coli* OP50 lawn and moist with 1 mL M9 solution or water.
- Incubate the plates for a few hours. During incubation motile nematodes crawl toward the bacterial lawn.
- Pick individual worms and transfer in new fresh bacterial culture media plates (both male and female species are required).

M. Senthilkumar, N. Amaresan and A. Sankaranarayanan, *Plant-Microbe Interactions: Laboratory Techniques*, Springer Protocols Handbooks, https://doi.org/10.1007/978-1-0716-1080-0_76,

1.4 Notes

- To avoid other bacterial and fungal growth moist with 0.01% thimerosal in M9 solution.
- Use autoclaved M9 solution or water.

2 Decanting and Sieving: Cobb's Method

2.1 Principle

The Cobb's method is suitable for the extraction of active nematodes from soil and sediments. This method is based on the nematode mobility, size, shape, and sedimentation rate between nematodes and soil particles. In this method, the nematodes are detached through sieving process. Heavy particles of soil settle down and the nematodes float on the upper layer, and the nematode suspension is decanted. Nematodes are extracted by using series of sieves of decreasing mesh size, so the different size of nematodes are collected separately. After a certain extraction the nematode suspension became clear suspension [2, 3].

2.2 Materials

- Soil sample.
- 2 L glass beaker.
- Three bowls (4 L capacity).
- A set of Cobb's sieve: 500–1000 μm, 350 μm, 175 μm, 100 μm, 45 μm.
- Stirring rod.
- Decanting tray with cross piece.
- Extraction sieve.
- Watch glass.
- Extraction dish.
- Clamping ring for securing the nematode filter.

2.3 Methodology

- Collect 100 g of soil sample and place in to beaker containing 1 L water.
- Stir well until homogenous suspension obtain, leave the suspension for 15 s then quickly decant the supernatant in bowl (repeat three times and discard the sediment).
- Pass the supernatant with the help of 500 or 1000 μm sieve in to another bowl and shake the sieve to help nematodes to pass through. Wash the debris from the sieve.
- Pass the suspension individually through 350 μm, 175 μm, 100 μm, and 45 μm sieves in the bowl and wash the debris from the sieve.
- Use clamping ring to attach the two nematode filters in the extraction sieve. Moist the filters to remove the air bubbles.

Place the sieve in water filled decanting tray with cross piece and place the watch glass on the filters.

- Pour the suspension from the collecting pan onto the watch glass on the filter. The last suspension should be mixed well. Remove the watch glass, lift the sieve and remove the clamping ring. Add a sample label to the filter.
- Place 100 mL water filled extraction tray in a vibration free area. Carefully place the well-drained sieve in the extraction tray (do not move the tray and cover it to prevent evaporation and dust).
- After 16–48 h extraction period, remove the sieve and nematode suspension pour in to a 100 mL beaker for analysis.

References

1. Barrière A, Félix MA (2006) Isolation of *C. elegans* and related nematodes. WormBook 17:1–9
2. Klass MR (1983) A method for the isolation of longevity mutants in the nematode *Caenorhabditis elegans* and initial results. Mech Ageing Dev 22:279–286
3. Van Bezooijen J (2006) Methods and techniques for nematology. Wageningen University, Wageningen, p 20

Chapter 77

Extraction of Nematodes from Plant Materials

Abstract

Plant parasitic and nonparasitic nematodes are present in leaves, stems, seeds, roots, tubers, and bulbs. Nematode extraction methods from plant materials are based on mobility of nematodes. Extraction methods include Funnel spray method, Dissection method, Blender nematode filter method, funnel spray method, and blender centrifugal flotation method. Dissection method is a direct method that can be directly used for microscopy with desired staining technique. Inactive nematodes can be isolated by the blender centrifugal flotation method.

Keywords Nematodes, Extraction, Funnel spray method, Dissection method, Blender nematode filter method, Blender centrifugal flotation method

1 Dissection Method

Dissection method is advisable for diagnostic purpose. By this method infected plant materials are analyzed under a microscope. Dissection method is also helpful to pick nematode egg mass [1].

1.1 Materials

- Infected leaf sample.
- Petri plate.
- Clean water.
- Forceps.
- Dissecting needle, handling fishing needle.
- Painting brush (no.1).

1.2 Methodology

- Wash the leaf sample and place in to the water filled in petri plate.
- Dissect the plant sample using dissecting needles and forceps using a 15–50× magnification.
- Pick the emerging nematodes and egg mass from the suspension with the help of handling needle or painting brush.
- After 2–3 h again observe the suspension, because some active stage nematode may crawl out of the material.

M. Senthilkumar, N. Amaresan and A. Sankaranarayanan, *Plant-Microbe Interactions: Laboratory Techniques*, Springer Protocols Handbooks, https://doi.org/10.1007/978-1-0716-1080-0_77,

2 Baermann Funnel

The Baermann funnel method is based on motility of nematodes that can be used for both plant and soil samples.

2.1 Materials

- Infected leaf sample.
- Clean water.
- Funnel with soft silicone tube attached and closed with squeezer clip.
- Pair of scissors or a knife.
- Cheesecloth.

2.2 Methodology

- Collect infected plant material wash it carefully and cut into 1 cm sized pieces.
- Collect subsample of 1 cm size and wrap it with a piece of cheesecloth forming loose ball.
- Take clean funnel and fill the clean water until it reach to 1 cm below the rim.
- Avoid bubble formation and make sure that the squeezer clip close.
- Hang the cheesecloth with sample in funnel for 16–72 h.
- Harvest the nematode suspension by opening the squeezer clip.

2.3 Notes

- Make sure sample never becomes dry.
- To avoid bacterial contamination add few mL of methyl-*p*-hydroxybenzoate (0.15%) in water.
- Use 0.15% solution of hydrogen peroxide instead of water.

3 Blender Centrifugal Flotation Method

3.1 Principle

This method is used for the extraction of both active and inactive nematodes and their egg mass from plant materials. This method is based on specific gravity. In extraction fluid, nematodes keep afloat and higher specific gravity fluid will sink. Nematodes can be collect though fine sieve.

3.2 Materials

- Infected plant materials.
- Domestic blender.
- Centrifuge.
- Pair of scissors.
- Vibro-mixer.

- 120 μm and 5 μm sieve.
- Centrifuge tube.
- $MgSO_4$ solution.

3.3 Methodology

- Wash the plant materials and cut in to 0.5 cm length. Place a subsample of a particular size in a domestic blender. Add water and blend for 5 s at 11,180 × *g*.
- Sieve the suspension with 1200 μm in to a plastic bowl. Rinse the residue on the sieve to collect the nematodes.
- Centrifuge the suspension at 1800 g for 4 min.
- After centrifugation, decant the suspension over a 10-μm sieve to collect nematodes which did not precipitate.
- Add $MgSO_4$ solution to the pellet and mix with a vibro-mixer for 20 s. Centrifuge for 3 min at 1800 g.
- Collect the supernatant and pour on the 10 μm sieves. Wash the sieve with water and collect the nematodes in a beaker.

3.4 Notes

- For egg mass collection 5 μm sieves required.
- $MgSO_4$ solution can be reused.

Reference

1. Van Bezooijen J (2006) Methods and techniques for nematology. Wageningen University, Wageningen, p 20

Chapter 78

Techniques for Maintaining Pure Nematode Population

Abstract

The techniques available for purifying the population of the nematodes in the laboratory includes—by growing on the solid media, nematode growth media, or by growing them in liquid media. These methods are used to maintain the nematode population in the pure form.

Keywords NGM (nematode growth media), Nematodes, Purification, Liquid media

1 Growth in Solid Media (NGM Media)

1.1 Principle

Nematode growth media (NGM) is a solid media used for the isolation of the nematodes. NGM agar plates are inoculated with *E. coli* and serve as a food source for nematodes [1].

1.2 Materials

- NGM plates.
- *E. coli* culture.
- Parafilm.
- Spreader.

1.3 Methodology

1. Prepare NGM plates before 2–3 days to use for detection of contamination.
2. Add approximately 0.05 mL (for medium size plate) or 0.1 mL (for large size plate) of *E. coli* culture on NGM medium plates and make a lawn in the center.
3. Allow to grow *E. coli* lawn at 37 °C for overnight.
4. After the incubation, inoculate the worms or nematodes and store at 11 or 16 °C for several months.
5. To prevent drying out plates wrap the plates with parafilm.

1.4 Notes

- Medium size (60 mm diameter) plates are useful for maintain general strain, large size plates (100 mm diameter) are useful for large quantities of worms.

M. Senthilkumar, N. Amaresan and A. Sankaranarayanan, *Plant-Microbe Interactions: Laboratory Techniques*, Springer Protocols Handbooks, https://doi.org/10.1007/978-1-0716-1080-0_78,

- Not to spread the lawn to the edge of the plate because worms may crawl up the side of the plate, dry out and die.

2 Growth in Liquid Media

2.1 Principle

Usually liquid culture of worms is grown on S medium using concentrated *E. coli* OP50 as a food source. In liquid media the worms are longer and thinner than those are grown on solid media and tend to hold their eggs.

2.2 Materials

- LB broth.
- Concentrated *E. coli* OP50.
- Pipettes.
- Reagents for S medium.
- Conical flask (250 mL and 1 L).
- Ice bucket.
- Centrifuge.
- Centrifuge tubes.

2.2.1 Preparation of Liquid Culture of Nematodes

Concentrated *E. coli* OP50: Grow *E. coli* OP50 overnight in LB broth or any other rich broth medium.

2.3 Note

- The concentrated pellet can be stored at 4 °C for several weeks or at −70 °C for indefinitely.

2.3.1 Preparation of Reagents

- S basal prepared with 5.85 g NaCl, 1 g K_2HPO_4, 6 g KH_2PO_4, and 1 mL cholesterol (5 mg/mL in ethanol) in 1 L distilled water and autoclave it.
- 1 M Potassium citrate (pH 6.0): add 20 g citric acid monohydrate and 293.5 g tri-potassium citrate monohydrate in 1 L distilled water. Then autoclave it.
- Prepare trace metals solution: 1.86 g disodium EDTA, 0.69 g $FeSO_4 \cdot 7H_2O$, 0.2 g $MnCl_2 \cdot 4H_2O$, 0.29 g $ZnSO_4 \cdot 7H_2O$, 0.025 g $CuSO_4 \cdot 5H_2O$ in 1 L distilled water and autoclave the mixture.
- Prepare 1 M $CaCl_2$: Add 55.5 g $CaCl_2$ in 1 L distilled water and autoclave it.
- **S media**: Take 1 L S basal, add 10 mL 1 M potassium citrate, 3 mL 1 M $MgSO_4$, 3 mL 1 M $CaCl_2$, and 10 mL trace metal solution (in aseptic condition).

2.4 Methodology

- Add 250 mL of S medium and concentrated *E. coli* OO50 pellet in 1 L flask.
- Take four large plates with nematode population and wash with 5 mL S medium to the 250 mL flask.
- Incubate the flask on shaking condition at 20 °C for 5–6 days.
- Monitor the culture by checking drop of culture under microscope.
- After incubation place the flask on ice for 15 min to allow the worms to settle.
- Aspirate most of the liquid from the flask and transfer remaining solution to the 50 mL sterile centrifuge tube.
- Centrifuge at 115 rpm for 2 min to pellet the worm. Young larva takes more than 2 min to pellet.
- Aspirate the resting liquid.

2.5 Notes

- Autoclaved trace metals solution to be placed in to dark condition.
- During monitoring the culture, if food supply is depleted then add more concentrated *E. coli* in S medium.
- Nematodes stocks can be maintained between 16 and 25 °C.

3 Transferring Worms Grown on NGM Plates

3.1 Principle

The chunking method is good for homozygous population and those have burrowed in to the agar. The filter paper method is used for only homozygous population.

3.2 Materials

1. **Chunking method**
   - Scalpel or spatula.
   - Nematodes culture plate.
   - NGM plates without worms.
2. **Transfer of worms by stripe method (filter paper)**
   - Filter paper.
   - Worms containing plates.

3.3 Methodology

1. **Chunking method**
   - Using sterilized scalpel or spatula take up chunk of agar from old plate and transfer to fresh plate. In chunk of agar about 100 of worms are present.
   - From the chunk, worms are crawl out on the new plate with bacterial lawn (*E. coli*).

2. **Transfer worms by stripe (filter paper)**
   - Take filter paper and cut into ½ to ¼ wide and 2–3 in. long size. Sterilize the filter paper and gently set on the nematodes containing plate.
   - The filter paper absorbs the moisture and picks the worms.
   - Deposit the nematodes on fresh NGM plate through placing the filter paper.
   - Then discard the filter paper after the use.

3.4 Note

- This method is not use for heterozygous population.

Reference

1. Stiernagle T (1999) Maintenance of *C. elegans*. *C. elegans* 2:51–67

Chapter 79

North Carolina Host Differentiation Test for Identification of Nematode Population

Abstract

North Carolina differential host test is the combination of resistant and susceptible reactions of six different hosts of the nematode population. This method is used to identify one of the four common species of *Meloidogyne* species. The North Carolina (NC) differential host test is based on the ability of nematode species to reproduce on a given host.

Keywords North Carolina differential host test, *Meloidogyne* species, Nematodes

1 Principle

The North Carolina (NC) differential host test [1] is based on the ability of nematode species to reproduce on a given host. This method is used to separate the *Meloidogyne* spp. (root-knot nematodes). The NC differential host test was successfully used to separate four main species of *Meloidogyne* such as *M. arenaria*, *M. incognita*, *M. javanica*, and *M. hapla* as well as host races of *M. incognita* and *M. arenaria*. The root-knot nematodes are the important groups of nematodes because they attack agricultural crops and cause economic loss [2, 3]. The NC host differential test is also used to characterize *Meloidogyne* spp. in conjunction with other species-specific morphological and biological markers.

2 Materials

- Pots.
- Sandy sterile soil.
- Glass staining dish.
- Nematodes egg mass (juveniles stage).
- Stereo microscope.

M. Senthilkumar, N. Amaresan and A. Sankaranarayanan, *Plant-Microbe Interactions: Laboratory Techniques*, Springer Protocols Handbooks, https://doi.org/10.1007/978-1-0716-1080-0_79,

3 Methodology

- Add 300 g of sterile sandy soil to the pot for the plant growth.
- Grow the test plants for the North Carolina host differentiation test (*Capsicum annum*, *S. lycoperscicum*, *Nicotiana tabacum*, and *Arachis hypogaea*).
- The collected nematodes egg mass hatch in tap water in glass staining dishes.
- Inoculate the thousand second stage juveniles to per pot.
- The experiment to be conducted in a growth chamber at 24 ± 1 °C.
- Provide daily 16 h photoperiod with watering the plant for 45 days. Conduct the experiment in triplicate.
- Prepare control with known amount *M. incognita* race 1 and *M. arenaria* race 2 populations.
- Remove the test plants root after 45 days then wash it and observe the gall formation under the stereo microscope and make galling index on a 0–10 scale.

4 Notes

- Produce single egg mass line of each population by inoculating *Solanum lycopersicum* cv.
- In galling forming index: 0 = no gall formation or fully developed egg masses containing juveniles.
- 1–4 = gall formation of secondary roots with fully developed egg masses containing juveniles.
- 5–10 = gall formation in tap root and primary lateral root.
- Scale 5 indicates that 50% of the test plant roots are galled and the population didn't fit in the current host race.

References

1. Hartman KM, Sasser JN (1985) Identification of Meloidogyne species on the basis of differential host test and perineal pattern morphology. In: Sasser JN, Carter CC (eds) an advanced treatise on Meloidogyne, vol 2. North Carolina State University, Raleigh, NC, pp 69–77
2. Robertson L, Diez-Rojo MA, López-Pérez JA, Buena AP, Escuer M, Cepero JL, Martínez C, Bello A (2009) New host races of *Meloidogyne arenaria*, *M. incognita*, and *M. javanica* from horticultural regions of Spain. Plant Dis 93:180–184
3. Stanton JM, O'Donnell WE (1998) Assessment of the North Carolina differential host test for identification of Australian populations of root-knot nematodes (*Meloidogyne* spp.). Australas Plant Pathol 27:104–111

Check for updates

Chapter 80

Isolation and Characterization Nematode-Egg Parasitic Fungi

Abstract

Nematodes are considered to be hidden enemy for the crops. However, various types of nematicides were used to control the nematodes but this was a short-term solution because the nematodes population increases after several months. Therefore, other approaches for nematodes management is the biological control of nematodes as it is feasible and eco-friendly method that will not allow the nematodes to develop its new biotypes. Recently, it has been investigated that egg-parasitic fungi is the potential biocontrol agent against nematodes.

Keywords Nematode, Egg-parasitic fungi, Nematicides, *Paecilomyces*, *Pochonia*

1 Principle

Paecilomyces and *Pochonia* are parasitic fungi of nematodes and they are known as pathogen of nematodes. Plant parasitic nematodes infect broad range of crops and alleviate the productivity. Around 150 species of fungi are known to attack nematodes or their eggs. There are few fungi available commercially for control of nematodes [1].

2 Materials

- Galled root.
- 1% sodium hypochlorite.
- PDA plates.
- Needle to remove egg mass from root.
- Inverted microscope.

M. Senthilkumar, N. Amaresan and A. Sankaranarayanan, *Plant-Microbe Interactions: Laboratory Techniques*, Springer Protocols Handbooks, https://doi.org/10.1007/978-1-0716-1080-0_80, © Springer Science+Business Media, LLC, part of Springer Nature 2021

3 Methodology

- Collect the infected galled root and wash with the running tap water.
- Carefully remove the nematode egg mass with the help of dissecting needle.
- Egg extracted from egg mass: Egg mass treated with 1% sodium hypochlorite for 1 min and wash the egg with sterile distilled water.
- About 100 eggs are spread on the PDA plate and incubate at 25 °C and routinely examine with inverted microscope (prepare five plates for one sample).
- Transfer growing fungal hyphae to the PDA plates for the isolation.
- Store the isolated egg parasitic fungi store at 4 °C.

4 Morphological Characterization of Nematode-Egg Parasitic Fungi

4.1 Principle

The lactophenol cotton blue (LPCB) wet mount method is widely used for staining and observation of fungi. The lactophenol cotton blue contains phenol which kills any live organism, lactic acid preserves fungal structure, and cotton blue stains the chitin in the fungal cell wall [2].

4.2 Materials

- Lactophenol cotton blue solution.
- MEA, YES, CYA media (for growth determination).
- Light microscope.
- Scanning electron microscope.
- 2.5% paraformaldehyde-glutaraldehyde.
- 0.05 M Phosphate buffer (pH 7.2).
- Cacodylate buffer.
- Isoamyl acetate.
- Ethanol.
- Fume hood.

4.3 Methodology

4.3.1 Light Microscopy

- Grow the fungi for the microscopic examination and growth rate determination.
- Prepare different media: malt extract agar (MEA), yeast extract sucrose agar (YES), czapeck yeast autolysate agar (CYA).
- Inoculate the fungi on different media plates and incubate at various temperatures for 7 days.

- Examine under light microscope using lactophenol cotton blue staining method.

4.3.2 Scanning Electron Microscopy

- Fix the culture in 2.5% paraformaldehyde-glutaraldehyde in 0.05 M Phosphate buffer (pH 7.2) for 1 min wash the sample with cacodylate buffer for 1 min.
- Again wash in cacodylate buffer and then dehydrating in graded ethanol and isoamyl acetate.
- Dry the sample in fume hood and mounted on to a SEM stub.
- Finally sputter-coated the sample with gold and observes under the scanning electron microscope.

References

1. Aminuzzaman FM, Xie HY, Duan WJ, Sun BD, Liu XZ (2013) Isolation of nematophagous fungi from eggs and females of *Meloidogyne* spp. and evaluation of their biological control potential. Biocontrol Sci Tech 23:170–182
2. Nguyen TTT, Paul NC, Lee HB (2016) Characterization of *Paecilomyces variotii* and *Talaromyces amestolkiae* in Korea based on the morphological characteristics and multigene phylogenetic analyses. Mycobiology 44:248–259

Chapter 81

Assessment of Nematicidal Activity of Microbial Volatiles

Abstract

Plant-parasitic nematodes are causing severe damage to the agricultural crops. Obligate root parasites such as root-knot nematodes infect more than 5000 plant species worldwide. To control these pathogens, chemical nematicides are used as they are efficient but it may cause serious environmental problems because of toxic residues they contain. Therefore, an alternative to chemical nematicide is the use of microorganisms which produces secondary metabolites (volatile organic compounds) for biological control of nematodes. These microorganisms are eco-friendly and release their volatile substances in the soil and fumigation of these volatile substances can help in prevention of various plant diseases. These volatile compounds when fumigated at certain temperature and pressure provide nematicidal activity against plant-parasitic nematodes.

Keywords Plant-parasitic nematodes, Chemical nematicides, Microbial volatiles

1 Principle

Three-compartment petri plate method is used to check nematicidal activity of microbial volatiles. In this method bacterial culture and live nematodes are placed respectively in a three-compartment petri plate and closed with parafilm to avoid escape of the volatile compound [1].

2 Materials

- Agricultural field soil.
- Beef extract peptone agar and beef extract peptone broth.
- Incubator with rotary shaker.
- Three-compartment petri plate.
- Charcoal.
- Live nematodes.
- Water agar.
- Parafilm.

M. Senthilkumar, N. Amaresan and A. Sankaranarayanan, *Plant-Microbe Interactions: Laboratory Techniques*, Springer Protocols Handbooks, https://doi.org/10.1007/978-1-0716-1080-0_81, © Springer Science+Business Media, LLC, part of Springer Nature 2021

3 Methodology

- Collect the agricultural field soil to isolate the bacteria.
- Isolate the bacteria using beef extract peptone agar.
- Inoculate the individual isolates in the BEPB (beef extract peptone broth) and incubate at 37 °C for 24 h on rotary shaker (200 rpm).
- Take three-compartment petri dish. Add 3 mL of the fresh culture of the isolate in one compartment about 3×10^{-7} per mL.
- Add water agar (WA) layer in other two compartments. Then inoculate the choice of live nematodes on the surface of WA (test to be carried out in triplicates).
- Immediately wrap the petri plates with the parafilm to prevent the escape of the volatiles and incubate the plates at 20 °C in dark condition for 24 h.
- For control, add same volume of BEPB instead of bacterial culture.
- Calculate the nematicidal activity (NA) using NA = IN/SN × 100%.
- IN represent the number of immobile nematodes and SN represents the sum of all nematodes counted.

3.1 Conformation Test

- Add active charcoal, bacterial culture, and nematode respectively in the three compartments of the petri plate.
- The activated charcoal is used to absorb bacterial volatiles. If the NA is indeed caused by the bacterial volatiles, expect no effect of the bacterial culture on the viability of the nematodes in the experiment.

3.2 Extraction of Volatile Compounds from Bacteria Using SPME (Solid-Phase Micro-extraction)

- Use 75 mm fibers supelco for SPME. First equilibrate with helium at 250 °C for 15 min.
- Add 9 mL of bacterial culture in 15 mL supelco SPME vials with 0.1 g sodium sulfate.
- Clamp the vials inside a thermostatic water bath and place on a hot stirrer.
- Equilibrate the sample for 1 h at 50 °C with constant magnetic stirring.
- Use 9 mL BEPB as a control.
- After extraction the SPME fiber is directly inserted into the front inlet of gas chromatography connected to a mass spectrometer and desorbed at 250 °C for 2 min.

- Identify the volatile compounds with similarity index from the database search based on a comparison of the mass spectrum of the substance with GC/MS system data bank.
- Each sample test is repeated twice.

Reference

1. Gu YQ, Mo MH, Zhou JP, Zou CS, Zhang KQ (2007) Evaluation and identification of potential organic nematicidal volatiles from soil bacteria. Soil Biol Biochem 39:2567–2575

Chapter 82

Trap Method for Screening Nematode Trapping Fungi

Abstract

In nematode-trapping fungi, conidial traps (CT) structures are formed directly on conidia germination without hyphal phase as an intermediate. These conidial traps structures are considered to be the survival structures which may overcome fungistasis in a drastic environment. CT structures are formed to increase the ability of the fungi in reducing the number of nematodes in cow dung environment.

Keywords Nematode-trapping fungi, Conidial traps, Conidia, Agar plate assay

1 Agar Plate Method

1.1 Principle

Nematode trapping fungi are formed by trapping structures of conidial traps (CT) through direct germination of conidia, without an intermediate hyphal phase. There are about 700 species of fungi which are able to attack living nematodes [1].

1.2 Materials

- Water agar plates.
- Rhizosphere soil sample.
- Freshly harvest conidial suspension (nematodes trapping fungi).
- Incubator.
- Inverted or conventional light microscope.

1.3 Methodology

- Prepare water agar plates (WA).
- Collect 5 g of water saturated rhizosphere soil spread on one-third surface on 9 cm WA plates.
- Preincubate the plates at 22 °C for 2–15 days before inoculation at start of the experiment.
- After preincubation period, inoculate the freshly harvested conidia suspension at the soil edge (position 1) and at 3 cm from the soil (position 2).
- Incubate the plates after conidia inoculation for 4 days.

M. Senthilkumar, N. Amaresan and A. Sankaranarayanan, *Plant-Microbe Interactions: Laboratory Techniques*, Springer Protocols Handbooks, https://doi.org/10.1007/978-1-0716-1080-0_82,

- After incubation observe the conidia traps (CT) using inverted or conventional light microscope.

1.4 Notes

- Prepare plates in duplicates.
- Fungi should be isolated from the agricultural soil for conidia preparation.
- Use freshly prepared conidia suspension.

Reference

1. Persmark L, Nordbring-Hertz B (1997) Conidial trap formation of nematode-trapping fungi in soil and soil extracts. FEMS Microbiol Ecol 22:313–323

Chapter 83

Direct Observation Method for Identifying *Pasteuria*: An Obligate Parasite of Nematodes

Abstract

Pasteuria species are found in soils having high pH and less organic matter. After adding baiting nematode, *Pasteuria* species are detected more in soils. *Pasteuria* produce compounds that are detrimental to plant-parasitic nematodes. This bacterial pathogen of nematodes is an assembly of various morphotypes and pathotypes. This bacterium infects various species of nematodes but does not infect organisms present in the soil and hence, considered as the obligate parasite of nematodes. When a contaminated nematode starts feeding on a root, *P. penetrans* spores germinate after few days and reproduces itself throughout the female body, and may either kill female or may mature but does not produce eggs.

Keywords *Pasteuria*, Baiting nematode, Obligate parasite, Nematodes

1 Principle

Pasteuria are endospore forming species and it has been investigated worldwide and demonstrated to play important roles in suppressive soil, especially root-knot nematodes. *Pasteuria ramosa metch* infecting cladoceran water Xeas, whereas, *P. thorneisher*, *P. penetrans,* and *P. nishizawae* are pathogens of root-knot nematodes. In the direct observation method, second stage juveniles (J2) naturally present in soil is extracted by baiting method [1].

2 Materials

- Materials for baiting method (see in the isolation part).
- 250 mL glass beaker.
- Aluminum foil.
- Inverted microscope.
- 6-well tissue culture plate.
- Sucrose for sucrose flotation method.
- Centrifuge and centrifuge tube.

M. Senthilkumar, N. Amaresan and A. Sankaranarayanan, *Plant-Microbe Interactions: Laboratory Techniques*, Springer Protocols Handbooks, https://doi.org/10.1007/978-1-0716-1080-0_83,

3 Methodology

- Extract the second stage juveniles (J2) from soil sample by baiting method.
- In this method additionally add J2 to increase the chance of detecting pathogenic bacteria and fungi.
- Add 3000 J2s on the surface of 50 cm^3 soil in 250 mL beaker, cover it with aluminum foil and incubate at room temperature for 2 weeks.
- After 2 weeks extract the J2s by sucrose flotation and centrifugation. Sucrose flotation method used for separate the organisms using their specific gravity.
- Transfer extracted nematodes to one well of a 6-well tissue culture plate and examine under inverted microscope (100–4000×) for infection.
- The first 100 J2s encounter examine for attached spore and infection.
- Distinguish *Pasteuria* sp. by morphology of the conidia attached to nematodes.

Reference

1. Ma R, Liu X, Jian H, Li S (2005) Detection of *Hirsutella* spp. and *Pasteuria* sp. parasitizing second-stage juveniles of *Heterodera glycines* in soybean fields in China. Biol Control 33:223–229